Political Theatre in Post-Thatcher Britain

Performance Interventions

Series Editors: **Elaine Aston**, University of Lancaster, and **Bryan Reynolds**, University of California, Irvine

Performance Interventions is a series of monographs and essay collections on theatre, performance, and visual culture that share an underlying commitment to the radical and political potential of the arts in our contemporary moment, or give consideration to performance and to visual culture from the past deemed crucial to a social and political present. *Performance Interventions* moves transversally across artistic and ideological boundaries to publish work that promotes dialogue between practitioners and academics, and interactions between performance communities, educational institutions, and academic disciplines.

Titles include:

Alan Ackerman and Martin Puchner (*editors*)
AGAINST THEATRE
Creative Destructions on the Modernist Stage

Elaine Aston and Geraldine Harris (*editors*)
FEMINIST FUTURES?
Theatre, Performance, Theory

Lynette Goddard
STAGING BLACK FEMINISMS
Identity, Politics, Performance

Leslie Hill and Helen Paris (*editors*)
PERFORMANCE AND PLACE

Amelia Howe Kritzer
POLITICAL THEATRE IN POST-THATCHER BRITAIN
New Writing: 1995–2005

Melissa Sihra (*editor*)
WOMEN IN IRISH DRAMA
A Century of Authorship and Representation

Performance Interventions
Series Standing Order ISBN 1-4039-4443-1 Hardback 1-4039-4444-X Paperback
(*outside North America only*)

You can receive future titles in this series as they are published by placing a standing order. Please contact your bookseller or, in case of difficulty, write to us at the address below with your name and address, the title of the series and the ISBN quoted above.

Customer Service Department, Macmillan Distribution Ltd, Houndmills, Basingstoke, Hampshire RG21 6XS, England

Political Theatre in Post-Thatcher Britain

New Writing: 1995–2005

Amelia Howe Kritzer

palgrave
macmillan

First published 2008 by
PALGRAVE MACMILLAN
Houndmills, Basingstoke, Hampshire RG21 6XS and
175 Fifth Avenue, New York, N.Y. 10010
Companies and representatives throughout the world

PALGRAVE MACMILLAN is the global academic imprint of the Palgrave Macmillan division of St. Martin's Press, LLC and of Palgrave Macmillan Ltd. Macmillan® is a registered trademark in the United States, United Kingdom and other countries. Palgrave is a registered trademark in the European Union and other countries.

ISBN-13: 978–1–4039–8829–4   hardback
ISBN-10: 1–4039–8829–3   hardback

This book is printed on paper suitable for recycling and made from fully managed and sustained forest sources. Logging, pulping and manufacturing processes are expected to conform to the environmental regulations of the country of origin.

A catalogue record for this book is available from the British Library.

Library of Congress Cataloging-in-Publication Data
Kritzer, Amelia Howe, 1947–
    Political theatre in post-Thatcher Britain : new writing, 1995–2005 /
Amelia Howe Kritzer.
        p.   cm. — (Performance interventions.)
    Includes index.
    ISBN 1–4039–8829–3 (alk. paper)
    1. Political plays, English—History and criticism.   2. Politics and
literature—Great Britain—History—20th century.   3. Politics and
literature—Great Britain—History—21st century.   4. Theater—Political
aspects—Great Britain—History—20th century.   5. Theater—Political
aspects—Great Britain—History—21st century.   I. Title.
PR739.P64K75 2008
822'.914093581—dc22
                                                                2007048566

Printed and bound in Great Britain by
CPI Antony Rowe, Chippenham and Eastbourne

Contents

Acknowledgements

I am grateful for various kinds of support, encouragement, and inspiration in the process of writing this book. I benefited from the comments of anonymous reviewers in the publication process.

Time and resources for the work were made available by research grants and a sabbatical leave from the University of St. Thomas.

With my husband Bert I have shared trips abroad, many nights of theatre, and lively political discussions. I acknowledge Bert's continuous encouragement of my work and dedicate this book to him.

1
Politics and Theatre

In a sense, all theatre is political. Theatre's context and referent is the world, and as John McGrath observed, 'There is no such thing as a de-politicized world' (2002: 199). While theatre is not the only art with political dimensions, it offers a unique forum for the political by involving audiences in a perceptible, if ephemeral, social reality through the operation of its conventions. Evidence of the close and perhaps intrinsic relationship between politics and theatre can be found in the long history of governmental regulation of theatre in degrees and forms that have not been applied to music, visual art, or written fiction. Theatre's most basic political potential lies in its paradigmatic relationship to the polis: within theatre's space, assembled citizens view and consider representations of their world enacted for them in the immediacy of live performance. As Richard Schechner states, drama is 'that art whose subject, structure, and action is social process' (121). Michael Kustow similarly describes theatre as both 'an art and ... a model of living together' (xv).

Writing of political processes, the philosopher Hannah Arendt argued that a necessary constituent of political freedom is a space in which freedom can be exercised – 'a place where people could come together' (25). Theatre provides such a space, in the most tangible sense, and the conventions associated with theatrical performance provide a rudimentary social organization of the performers and spectators. Though its free status is always mediated by multiple economic and regulatory factors, theatre offers a medium for exposing problems, exploring issues, advocating action in public or private life, and experimenting with changed relations of power within the context of a form that participates in the social in a variety of direct and metaphoric ways.

Theatre's particular power consists not only in the space or the audience assembled there, or even in its representations *per se*, but also in the

1

means of enactment – actors playing roles. The psychological interplay between the real actor and the fictional role gives theatre the characteristic that Victor Turner has termed 'liminality', to indicate location in a border zone of experience and consciousness. Schechner and other scholars have pointed out that theatre shares the quality of liminality with religious and social rituals involved in the formation and maintenance of individual and group identity. Erica Fischer-Lichte argues that theatre's central function, occurring through the interactions between actor and role and between actor and audience, is the staging of identity. Fischer-Lichte suggests that the human 'finds himself via the detour of another' and that theatre 'symbolizes the human condition of creating identity to the extent to which it makes the distancing of man from himself the condition of its existence' (3, 5). Bringing together the theories of Helmut Plessner, Denis Diderot, and Judith Butler, Fischer-Lichte postulates that 'the actor's skill in staging the identity of a role . . . allows [the spectator] to play with different identities and possibly even encourages him to make a change in his identity' (4). The relationship between theatre and identity traced by Fischer-Lichte in her historical analysis of European drama asserts a remarkable potential of theatre as a form of political expression. Identity structures political understanding, choice, and action. Changes in, and even heightened awareness of personal or group identity lead to changes in perception of the body politic and all that relates to it.

Contemporary theatre couples this powerful psychological interaction with a freedom often taken for granted but none the less remarkable. In Western democratic societies, theatre may deal with any subject brought to its stages. Because theatre attendance is a voluntary and conscious act, theatre is not currently subject to the restrictions placed on television and even film – forms that to a greater extent inhabit what can be deemed public environments. Taboo-breaking plays created controversy and heightened interest in theatre periodically throughout the twentieth century, from the utterance of obscenities in Alfred Jarry's *Ubu Roi* to the explicit presentation of homosexuality in Tony Kushner's *Angels in America*. Even in times and places where theatre has been restricted, the multiple meanings inherent in the semiotics of performance have given theatre the means by which to initiate forbidden discourses, such as in the staging of Jean Anouilh's *Antigone* in Nazi-occupied Paris. The freedom of the theatre, moreover, goes beyond the ostensible topic of the play or performance presented. Theatre is always about form as well as topic, and performance has the potential to destabilize definitions and identities. The cross-gender roles in Caryl Churchill's *Cloud Nine*, for example, stimulated profound questioning and reorientation regarding gender identities.

Despite theatre's political potential, relatively few plays of any era exhibit overtly political aims. Those that do must overcome both artistic and social challenges to find and reach audiences. First among these challenges is the audience's desire for a pleasurable escape from the world around them. Escape in some form is inherent in the art of theatre, as actors take on the identity of characters and create an artificial world on stage, while audience members invest mental and emotional attention in these characters and their world. Pleasurable escape, however, implies distance from problems or situations that call for concrete action. Even ancient Greek audiences may not have appreciated seeing situations that were too close to current reality, since Aeschylus's early play *The Persians* is the only Greek tragedy known to dramatize a contemporary historical event. Daphna Ben Chaim, building upon the work of Edward Bullough and Jean-Paul Sartre, theorizes that psychical distance, or the ability to visualize the art object as a fiction, is necessary to the experience of the aesthetic. In a practical sense, the theatre experience generally implies a brief holiday from responsibility, with no intrusive reminders of actual obligations. Of course, politics itself can provide such a holiday for some audience members, especially when its entertainment value is heightened by personal misconduct, as in the recent tabloid-churning adventures of former British Home Secretary David Blunkett. However, it should be obvious that distance from the issues in question heightens their entertainment value; one might reasonably speculate that the prime minister, for example, would not consider misconduct on the part of one of his cabinet members pleasurably entertaining.

Subtler in its operation, a commonly perceived separation of cultural institutions from political life poses a challenge to political art. This separation arises from a model of culture as a quasi-sacred sphere dedicated to the preservation of works that exemplify truth and beauty, uncorrupted by political discussion or topical issues more generally. Political or even topical content stands as a mark of inferior work that should be excluded from the cultural sphere. This model of culture centres on veneration of the past and established masters, as well as a philosophical belief in universally understood standards of truth and beauty. It creates an elite group of educated individuals with the power to select and interpret cultural presentations. While the idea of culture as a pure font of truth and beauty might appear ridiculously outmoded in the contemporary intellectual context, it has by no means been laid to rest, despite the best efforts of modern and postmodern criticism. Control of the cultural sphere by a specialized and non-democratic elite remains unchanged. Postmodern critiques have actually reinforced boundaries of cultural authority through the elite nature of their discourse.

Many plays included within the canon of world drama contest the ostensible boundary between the cultural and the political. Political theatre in Britain draws upon the strengths of a rich tradition that goes back at least to Norton and Sackville's *Gorbuduc*, in 1562. The Elizabethan theatre played a role in the politics of its time (see, e.g., Mullaney's 1995 study), and the plays of Shakespeare contain a strong current of political commentary (see, e.g., Dollimore, 2003). Theatre became a symbol of royal decadence in the Commonwealth period, and an arena for expression of expanded social freedom in the Restoration. The early eighteenth century produced a vigorously oppositional political theatre; its anti-government satires brought on the Licensing Act of 1737. Late Victorian theatre included activist writers such as G.B. Shaw, John Galsworthy, Arthur Wing Pinero, and Somerset Maugham, with dramas that commented on class, gender, and other socio-political issues. Plays deeply involved with the Irish nationalist movement, such as those by J.M. Synge, Augusta Gregory, and Sean O'Casey, have formed the core of an enduring political consciousness. The early-twentieth-century workers' movement produced overtly political plays (see Goorney and MacColl, 1986), in spite of the fact that the Lord Chamberlain's office could withhold permission for the performance of plays considered politically objectionable.[1]

More recent developments provide the backdrop against which the political expression of contemporary plays must inevitably be discussed. British political theatre acquired a unifying identity and international visibility in the late 1950s and early 1960s, when the voices of young men with working-class origins were heard for the first time. The 'angry young men' rebelled against the proverbial patience and self-effacement of the English common people. Their anger can be viewed in the light of what later came to be known as identity politics. In demanding a place in the economic and social structure of the middle class for those not born into it, the angry young men implicitly sought changes in the organization of a tradition-bound, class-based society. Although recent reassessments of the plays of this era have offered alternate interpretations of their political significance (see Shellard, 1999; Rebellato, 1999), the unifying idea of anger brought socialist playwrights such as Edward Bond, John Arden, and Harold Pinter to public attention and created an alliance between theatre and leftist politics.

The end of official censorship of theatre in 1968 brought greater freedom of expression and a diverse range of demands for recognition and power. Activists who organized around such issues as nuclear disarmament, feminism, gay rights, and regional identity contributed to an explosive increase in the number and variety of politically oriented theatres. The agendas of political theatre expanded beyond trade unions, socialism, and the working class, to engage with issues of race, ethnicity, gender, and sexuality. Companies with varied political agendas – including Foco Novo, Freehold, Portable Theatre, Belt and Braces, Gay Sweatshop, Joint Stock, Monstrous Regiment, the Tara Arts Group, Northwest Spanner, Red Ladder, the 7:84 Companies in England and Scotland, Cartoon Archetypal Slogan Theatre (CAST), Women's Theatre Group, and Welfare State International – were established during the 1970s.[2] At the same time, a number of established theatres in London and other cities added small studio spaces primarily for in-house production of new plays. The small theatres, art centres, and touring companies of the alternative theatre movement began to be referred to as the Fringe – a term that in itself carried obvious political implications.[3]

As its scope and aims expanded beyond specific issues to a broader expression of hopes for a less destructive and more egalitarian, spontaneous, and pleasure-based society, political theatre became a counter-cultural movement known as alternative theatre. Drawing on a reservoir of idealism and communitarian energy, the alternative theatre movement envisioned a refashioning of the cultural landscape. Thus the work of alternative companies often made form part of the message. They experimented with new techniques of creation, performed in community centres, pubs, schools, village halls, clubs, and on the streets, as well as in smaller theatres, and favoured what Michael Patterson calls 'interventionist' strategies over 'reflectionist' ones such as realism. The pervasive spirit of rebellion found expression in the National Theatre's first season in 1976, which included *Weapons of Happiness*, a celebration of youthful anarchy written by Howard Brenton and directed by David Hare. In their survey of recent British theatre, Richard Eyre and Nicholas Wright observe, 'Not every playwright of the '70s was politically on the left, but being left-wing was the mood of the time' (2001: 281). Government arts subsidies were a major factor enabling the development of alternative and regional theatre in Britain in the 1970s, as the Arts Council increased its level of funding under the Wilson governments (1964–70; 1974–76).[4]

When British politics made an abrupt shift to the right in the 1980s, the alternative theatre culture that had risen as a jubilant wave of exploration and experimentation contracted to a position of opposition, while the era of Conservative Party leadership changed both material and psychological dimensions of the cultural landscape. As Thatcher's policies gained a stranglehold on the Arts Council and other institutions, such as borough councils, that supported political theatre, funding cuts and outright elimination of subsidies drove many smaller theatre companies – especially those most associated with leftist politics – out of existence. Higher costs coupled with new requirements for obtaining corporate sponsorship and making detailed financial reports strained the resources of all subsidized theatres, exhausted the energy of many small companies, and reduced their productive capacity.

As theatres struggled to survive, the playwrights who expressed opposition to Thatcherism maintained the philosophical underpinnings of the political theatre of the 1970s, even as the tide of optimistic idealism ebbed. Their attacks targeted specific policies, directions, and effects of the Thatcher government. Howard Brenton and David Hare satirized the takeover of a large sector of British journalism by a wealth-obsessed entrepreneur who embodied Thatcher's free market ideals in *Pravda* (1985). Caryl Churchill sharply interrogated the assumptions and values of Thatcherism in *Top Girls* (1982) and *Serious Money* (1987). Jim Cartwright's *Road* (1986) and Kay Adshead's *Thatcher's Women* (1987) dramatized conditions in the north and west of England, where entire industries had collapsed and millions of workers were unemployed. Increasingly through the Thatcher era, as Peacock has pointed out, playwrights retreated into explorations of personal, rather than public, issues, while new theatre companies tended to direct their energy towards aesthetic experimentation rather than social or political goals. By the early 1990s, the energy of opposition had dissipated, and political theatre, no less than political parties of the left, had failed to articulate ideas that expressed the aspirations of the majority of the public.

Recent trends in Britain have created new challenges for political theatre. Politics has entered a less oppositional phase. The mid-1990s brought electoral rejection of Conservative politics but no renaissance of the political left. The Labour Party, which failed to oust the Tories until it recast itself as New Labour, took power by discarding socialist goals and the grass-roots passion for political wrangling with which Labour had long been identified. In the post-Thatcher era, the public shows a trend towards decreased identification with ideologies, as well as a tendency to view with scepticism the claims of political parties and organizations to offer solutions to societal problems.[5] Divisive issues such as animal rights and the Israeli–Palestinian conflict stand in the way of re-establishing a broad-ranging leftist coalition. The end of Thatcher's government also did not bring about a clear reversal of the policies she instituted. New Labour has in some ways accommodated to Thatcher's legacy, and the Conservative Party has also moved towards the centre on issues such as affiliation with the European Economic Community.

Political convergence and the breakdown of ideological identity have reoriented political theatre. Drama relies on conflict and proves most readily understood if the parties to the conflict are easily identifiable. Political drama in Britain today cannot rely upon familiar oppositions; instead, it must structure and define a political landscape, before it can stake out positions. Even then, a dramatist cannot be sure that audiences will recognize the landscape or understand the positions. As Max Stafford-Clark asserted in a discussion reported by Michael Billington, 'In the '80s we all knew who the enemy was. Now we are not so sure' (2001). The current political environment, dominated by the deliberately vague surface of New Labour politics, has created an ideological vacuum that serves to disable activism and foster cynicism. The political theatre of the 1970s continues to exert power as a model of political drama; however, the energy and optimism that fuelled it has dissipated.

The re-emergence of political theatre in the late 1990s thus came as a surprise to critics and audiences. In 2003, the arts editor of the *Guardian* decided to explore this phenomenon by asking currently active playwrights about their interest in and practice of political theatre. The *Guardian* launched a series of essays in which playwrights responded to the question 'What is political theatre?' A remarkable selection of dramatists offered a spectrum of definitions and defining strategies. Two veteran playwrights argued against explicitly political plays. Pam Gems insisted that politics 'belong on the platform, in the committee room, on the march' (17 May 2003). Arnold Wesker asserted, 'My own work begins with human beings, not ideas' (*Guardian* 15 March 2003). Younger playwrights took a more complex view of political theatre but acknowledged disadvantages in being tagged as a political playwright. Michael Wynne asked if plays 'deemed political . . . will end up like the ballot box with no one interested, no one coming?' (3 May 2003). Gregory Burke stated that in his play *Gagarin Way*, 'there is a lot of talk about politics, but what it is really about is community' (12 April 2003). Even David Edgar qualified the political aspect of his work by highlighting his fascination with the 'human passions' that underlie politics (19 April 2003).

Writers who personally and artistically represent ethnic or racial minorities in Britain argued that, for them, politics are unavoidable. Gary

Mitchell explained that actors have approached his plays about working-class Loyalists in Ulster on the basis of whether they portray Protestants in a negative light.[6] Kwame Kwei-Armah wrote forcefully of the challenges in working within the mainstream theatre, with its pressure for assimilation, while maintaining a distinctively Afro-Caribbean voice and viewpoint. The convergence of the personal and political became clear for him in the repeated experience of having strangers commend and thank him for marrying a black woman, in contrast to the pattern of many successful black men. Rather than seeking to evade it, Kwei-Armah embraced the political, defining his political goal as using the stage to 'refract through the humanity of my cultural lenses' (10 May 2003).

A pair of playwrights separated by a generation located their understandings of political theatre within a historical context. Mark Ravenhill argued that Anglo-Irish drama has long been defined by a close association with politics. He placed his own work within a 'pragmatic, materialist' tradition that focuses on the 'sociological, the anthropological, the political' (22 March 2003). With ironic humour, Ravenhill suggested that government subsidy of theatre perhaps obligates contemporary playwrights to justify the public's investment by contributing to an understanding of society. Nevertheless, he concluded, 'to capture the truth of the new world we live in is an exciting ambition' (*ibid.*). David Hare offered a personal historical perspective in a moving evocation of what it has meant to 'choose drama' as a means of political expression:

> Failure is our element. Theatre has changed as little as society. Yet many of us have ended up curiously buoyant, not, let's hope, consoled but braced by the beauty of what we are attempting, in art as much as in politics. We are sustained by the thing itself, its superb difficulty … it is remarkable how many of us feel that even if it has been a lifetime of failure, it has not been a lifetime of waste' (24 May 2003).

Analysis of the interaction between theatre and politics in Britain today, which is the aim of this study, must start by answering the question posed in the *Guardian* essays. This response to the question 'What is political theatre?' will begin with a discussion of what is meant by the political. Politics, the set of processes through which power is exchanged, structures social systems and operates at every level within a society. The ideal of liberal democracy envisions a broadly inclusive political system with equitable representation of groups and classes and peaceful resolution of conflicts between different constituencies. The incomplete realization of this ideal has formed the basis for most twentieth-century political movements in Britain and other democracies. More recently, the ideal itself and its underlying assumptions have been questioned by postmodern theorists.

Although politics pervades social life, only a small part of its operation draws public attention. Elected officials, political parties, and governing bodies such as parliament or local councils constitute what is most often perceived as the political sphere. Such public-sector institutions as education and health care are generally understood to have a political aspect. Beyond these obvious manifestations, much of politics remains hidden unless an interested party makes it visible – in other words, makes an issue of it. Political theorist Murray Edelman has written about the visible political sphere as 'the political spectacle' (1988), using a consciously theatrical term to describe phenomena that are widely perceived as political. Making particular issues or actions visible constitutes the fundamental operation of political activism. Edelman characterizes political developments as 'creations of the publics concerned with them' (1988: 2). Obviously, the more powerful a particular group within a society, the greater control it exercises over the visibility of issues and actions. Furthermore, given the central role played by the mass media in constructing the political spectacle in contemporary societies, those with greater access to the media by definition exercise greater political power.

Visibility is one path to power within a political system. A group, or even a single individual under some circumstances, may construct an issue through some type of public campaign. If successful, the campaign will capture media attention, provoke further campaigning, and eventually result in bureaucratic and/or legislative action. Interpretation provides a second, equally important means of gaining power. As Edelman observes, 'in every era and every national culture, political controversy and maneuver have hinged upon conflicting interpretations of current actions and developments' (2). Thus, an interested party may construct a reinterpretation of an already visible issue, initiate a campaign to promote the alternate interpretation, and seek through this campaign to convince others of that view.

A political system offers endless potential issues and interpretations. The achievement of visibility means being perceived in distinction to a mass of competing items in an array of information sources. Considering that most sources of news and information are commercial ones, the competition arises not only from different news stories but also from advertising. Reading a newspaper, one may find one's sympathies aroused by a report of suffering in a famine, be moved to moral outrage by an article about the torture of prisoners, but actually take action to purchase something in response to an advertisement. Given the constant and varied flow of information, any issue must fight a Darwinian battle to make it to the forefront of public consciousness. Once there, it faces the even greater challenge of holding the attention of the public for the length of time necessary to accomplish any political change. The confusion and distraction generated by the multiplicity of issues, constant streams of information, and persuasive appeals of ubiquitous advertisements play a part in generating apathy and leading to political disengagement. Disengagement, which Edelman terms 'the power of indifference' (8), constitutes a factor that is far more influential than activism. If, therefore, an issue threatens entrenched interests, those interests may oppose the issue not only through direct means but also by relying on the power of indifference. Indifference, in fact, plays a crucial role in keeping most of politics out of the public view entirely.

Visibility, moreover, cannot constitute an end in itself. Once perceived as political, an issue by definition becomes available for competing interpretations. An initial victory for those seeking visibility can be turned around if a competing interpretation gains ascendancy. Thus, for example, the feminist movement that achieved visibility in the United States in the 1970s by defining itself as a means of liberation was later reinterpreted, with devastating effectiveness, by a prominent right-wing talk show personality as an authoritarian ideology comparable to Nazism.[7] Issues and goals that involve disrupting or working outside established patterns of political action are especially difficult to achieve, even when they capture attention in a dramatic manner, because they are most vulnerable to oppositional reinterpretation.

The contemporary plays and performances described as political in this study are those which attempt to create political meaning by making visible and/or interpreting particular social phenomena as public problems or issues. In other words, theatre is considered political if it presents or constructs a political issue or comments on what is already perceived as a political issue. Defined in this way, political theatre initiates a dialogue with the audience about politics within a national or cultural system shared by both the creators of the theatre production and the audience. Political theatre often addresses its audience explicitly, but its means of communication extend beyond such obvious tactics and beyond the text, encompassing such things as the style of performance, the layout of the performance area, the use (or non-use) of theatre technology, the location of the theatre, the identities of the various people and institutions involved in the production, and the information provided by programs and publicity.

The audience, of course, though conventionally silent (except for polite laughter) in contemporary theatre, is not merely the passive recipient of a play's message. Audience presence defines live theatre (see Grotowski, 1975). The audience exercises an essential function in the creation of meaning in theatre. Its responses feed back into performance in a variety of powerful ways. A specific audience may create, either through outright organization or selective attendance, a theatre or set of theatres addressing it specifically. Theatre is not a monologue; it is a dialogue, though in most contemporary theatres the audience does not speak. Walter Benjamin's description of the way in which meaning is created through dialogue in *The Conversation* (1913–1914) applies particularly to theatre: 'the listener is really the silent partner. The speaker receives meaning from him; the silent one is the unappropriated source of meaning' (*Selected Writings*, I: 6). The power of the audience lies in its freedom to attend or not, to remain or not, and to create interpretations of the production. As Susan Bennett points out in her 1990 study, the participation of audiences in the theatrical dialogue is complex, framed by expectations about the way in which it will occur that have been formed through previous theatre attendance and various kinds of formal and informal orientation to theatre practices. A performance may reinforce these expectations or challenge them, and this choice affects the political dynamics of the dialogue.

The major debates about political theatre, and the various forms of drama and performance that have come out of those debates, focus on different ways of engaging the audience in dialogue and shaping audience reception. Clearly, theatre's capacity for political communication encompasses all the means through which it organizes the audience's experience and produces the theatrical spectacle and the performance. The audience brings to the theatre a willingness to involve themselves in a reality different from their usual everyday life, an expectation that the production will give them an opportunity to create meaning for themselves through interpretation, and a capacity to join their imaginations in the joint enterprise of creating this theatrical reality. Political theatre, characterized by a 'transgressive alterity' (Murray, 1997), creates a collision with existing assumptions, ideas, or perceptions about society and politics. This collision, which may occur by means of any combination of dramatic content, performance style, and theatrical context, serves as the initiating mechanism of the dialogue between production and audience. This collision must be created in partnership with, not in opposition to, the audience. To bring about a change in thinking, the theatrical dialogue must stimulate internal dialogues in which audience members themselves use the new perceptions and ideas made available in the performance to challenge perceptions and ideas they had previously recognized or accepted. Unlike the scripted dialogue of the stage play, the internal dialogue of any given audience member remains open-ended, moving towards change in ways and towards conclusions that cannot be scripted or even completely predicted. Benjamin poetically evokes this action: 'The speaker immerses the memory of his strength in words and seeks forms in which the listener can reveal himself. For the speaker speaks in order to let himself be converted' (*ibid.*).

Benjamin's suggestion runs counter to the stereotyped concept of political theatre as a polemic that demands agreement or disagreement. However, to the extent that political theatre provokes thought, interpretation, and discussion, it necessarily invites challenge and revision to the issues and interpretations it presents, as well as to the style and context of its productions. On a more fundamental level, the idea that speaking expresses the desire for conversion relates to the quest for identity that Fischer-Lichte posits as the basis of theatre. Conversion is more powerful than persuasion; it implies not just a change of mind, but a change of identity. In the context of theatre, the speaker – the playwright perhaps, or the ensemble of actors playing roles – seeks identity through the perceptions and interpretative acts of audiences. At the same time, the audience itself assumes a mask while in the theatre, performing an illusion of unity through simultaneous presence and silence. The conclusion of the performance breaks the audience's illusion of unity, as they resume their individual identities. The detour created by laying aside individual identity offers the audience member the power to interpret the speaker – assign it an identity. The identity shifts inherent in this almost paradoxical interplay of passivity and power both complicate and intensify the creation of political meanings in theatre.

Theatre that, for the purposes of this study, is not defined as political does not arouse its audience to an awareness of the political. Its created realities exclude social phenomena perceived to be political or controversial. Its subjects and forms do not invite comparison with political power, challenge existing systems of power, or provoke reinterpretation. Of course, as the opening quotation from John McGrath implies, even apparently non-political theatre can and often has functioned politically within a society. The Roman emperors, for example, with their 'bread and circuses' policy, seem to have understood that drama with emphatically non-political content, such as the domestic farce, could contribute to the maintenance of power by representing opposition and conflict as ridiculous. Theatre that does not attempt to draw its audience into a dialogue about politics may serve the existing power structure by contributing to apathy and disengagement. While this study does not focus on plays, performances, or theatres that lie outside its definition of political, it is important to remember that all theatre serves some political purpose, and that this purpose can usually be identified if the question is asked.

Theatre interacts with politics in part by imitating its operation within a society. This imitation is most apparent in the theatrical environment and conventions, such as the positions from which actors customarily perform. Theatres and performances create sets of social dynamics, exercising power over a group of people within the boundaries of a particular time and place. The structure of the particular theatre or production determines where the performance occurs, who is admitted and on what terms. It determines not only which subjects are addressed and which excluded, but also to what extent the process and means of selecting and presenting the subject is revealed. Different types of theatre exercise different degrees and forms of power. One may draw a clear contrast, for example, between types of interaction with the public created by a traditional theatre located in a bounded space and admitting only ticket-holders, and a street-theatre performance, which intervenes in the social life of passers-by. The audiences for these different types of theatre constitute very different social/political entities. The traditional theatre audience has entered into a kind of contract, accepting the constraints of selective admission, space, and convention in return for the right to experience the performance. The audience for street theatre finds itself unexpectedly caught up in the event, whether they wish to be or not, but may exercise their freedom to disrupt the performance or walk away from it with no loss of money or breaking of an implicit contract.

The form of interaction, no less than issues of access and cost, has often moved politically oriented theatrical productions towards the less formal and more socially integrated arenas of street theatre, pub theatre, or storefront theatre. Such productions, as Susan Bennett argues, may simultaneously challenge both the structured understandings of everyday life and the cultural 'frame' created by the audience's expectations of theatre (p. 3). In addition, they may serve to equalize power among audience members, cancelling the advantage enjoyed by the more experienced and knowledgeable. Even more significantly, performances of street theatre create a sense of revolutionary movement as they confront the status quo and engage the attention of random members of the public, offering them privileges of spectatorship that they had neither sought nor expected. Productions located in traditional theatre structures and employing the convention of a silent audience in a fixed location may serve to reinforce existing power relations within the audience; furthermore, their structure does not challenge the theatrical framing mechanism or, by extension, the basic underpinnings of society. At the same time, a production located in a traditional theatre may be able to make a stronger impact on an audience as a result of resources available in its specialized space (i.e., set, lighting, soundproofing) and because certain forms of expression (i.e., nudity, obscene language, simulated gunplay) not permitted in public spaces are generally protected within the bounds of legitimate theatre.

Even at its most mainstream and serious, theatre does not occupy a position of power in Britain or other contemporary societies. It exists outside the central structures of government and industry, and occupies only a minor place in the institutions of business, education, and religion. Theatre's position within contemporary culture is marginal compared to commercial television and film.[8] On one hand, theatre's marginality lowers barriers to participation. Furthermore, its perceived unimportance, coupled with the understanding that its presentations lie in the realm of illusion, allows it a degree of freedom not evident in official productions (such as televised press conferences by political leaders, parliamentary debates, or political party meetings), television programmes, or commercial films. On the other hand, its marginality limits reception: street theatre or other informal venues reach only those people in the immediate environments, while the audience for traditional theatre is limited to those with sufficient resources who consciously choose to attend theatre.[9] Theatres and producing companies generally depend on a variable combination of ticket sales, public funding, grants from charitable foundations, and corporate sponsorships. Competition for the support selectively given by such funding sources contributes to the relatively weak political position of theatres.

This set of circumstances highlights the problem long acknowledged by practitioners of political theatre: genuine impact is rare, and the process to it uncertain. The challenge becomes even more difficult when a company or production seeks to change minds, rather than to provide a sense of solidarity for audience members already in agreement with the position or cause being expressed. Theatre cannot compel change; it can only attempt to inform, provoke thought, change attitudes, and perhaps point the way towards action. Works with overtly political messages may fail to attract individuals or groups not already sympathetic to their point of view. Plays with political messages that are subtle, covert, or implicit may not be clearly understood. In either case, the work's impact may be slight or non-existent. Producing organizations and dramatic works most often attempt to negotiate this set of contradictions by combining the appeal of novelty, elements of entertainment, and a message that is persuasive rather than dogmatic and often coded or indirect. The multi-dimensionality of such works diffuses their political intent and impact. Occasionally a production tactically or inadvertently confronts the audience's expectations in such a way that some will walk out. Their action breaks the mask of the audience, brings a reassertion of individual engagement on the part of those who remain, and intensifies the experience of producing meaning for those audience members. The failure of political theatre, as in any type of production, occurs when audiences remain but cease to engage with the performance.

This is not to say that political impact is impossible to attain. From the beginnings of modernism in the late nineteenth century, theatre has been used as a forum for making social phenomena visible and, in some cases, defining them as political issues. Notable examples of success have created models of effective strategy. Gerhard Hauptmann's *The Weavers* (1892), a watershed political play, took as its subject the 1844 revolt by cloth weavers in the Silesian district of Germany. It used the then-new style of naturalism to expose the working conditions of poor workers. The extent to which *The Weavers* collided with dramatic expectations conditioned by romanticism and melodrama can be discerned from the commentary of a contemporary critic, Edward Everett Hale:

> In 'Die Weber' [Hauptmann] goes as far as one can readily imagine the stage can go. [The play does not take] the weavers' strike for a background, or a setting, or a situation.... The play takes the strike itself for its subject. There is no hero and no heroine; characters there are, but only because there must be people on stage to have any play at all (40).

The model provided by Hauptmann attempts to create an accurate and compelling representation of a social injustice. This model still demonstrates its viability.

The Weavers also provides the starting point for what Neil Blackadder has termed 'the theatre of opposition'. Blackadder focuses on the response to late-nineteenth- and early-twentieth-century plays that 'confronted the audience with depictions of human and social relations, and approaches to theatrical representation, which forcefully challenged conventional thinking' (14). Although, as Blackadder points out, theatre norms had recently evolved in the direction of enforcing silence, these plays provoked such profoundly strong opposing reactions that audience members actually performed their sentiments for and against the particular play, in the form of protests and riots in the theatre. Confrontational tactics in the plays of this era begin with naturalistic and realistic styles, which involve partial nudity and frank language, coupled with explicit representation of socially taboo subjects, such as childbirth in Hauptmann's *Before Sunrise* (1889). The shock value of the earliest of these plays depends on making visible situations and behaviour that were hidden and denied in bourgeois society.

The tactics of confrontational theatre have evolved over time. Social realists like Henrik Ibsen and George Bernard Shaw tempered the confrontational nature of their plays by placing the confrontation within the action of the drama and using sympathetic characters to interpret situations. Shocking material, such as inherited syphilis in the son of a respected man, is filtered through an unambiguous moral lens via characters involved in the action, rather than being presented directly to the audience. Both Ibsen and Shaw gained their reputation for confrontation by setting up moral positions in opposition to the conventional morality and religion of their day. They highlighted the inadequacy of these institutions for channelling social energies to benefit weaker members of society; however, they did not directly accuse audience members of hypocrisy or moral weakness. Agit-prop plays developed out of the workers' movements of the early twentieth century, as the lengthy and indirect methods of social realism did not serve the movement's need for brief, exciting, and easily produced dramatic pieces. Agit-prop tactics arouse the audience to a unanimous sentiment of opposition towards a third party with power or authority. Further developments of theatrical confrontation include challenges to accepted language and traditional structures of plot, character, and visual design, as well as transgressive use of national, religious, and cultural symbols.

Confrontational plays force the audience into a binary choice of acceptance or rejection. Their value lies in the directness and strength with which they contest the status quo. They are most important in bringing an issue to visibility. The vehement emotions they evoke tend to preclude complex interpretation – at least, not during their period of initial public exposure. Rejection may mean closure of the mind or leaving the performance, while acceptance may derive from the desire to be part of a rebellious faction. Though often successful in gaining visibility for an issue, shock tactics gradually lose their power as the audience becomes accustomed to whatever originally provoked outrage. As is clear in contemporary language on stage and off, what generates shock in one generation may be taken as the norm by the following generation.

Bertolt Brecht, whose first plays were of the confrontational type, attempted to engage the audience in ways that would provoke inner dialogue rather than merely approval of or disagreement with a stated message. Building on the work of Erwin Piscator, Brecht sought to create a theatre that rejects a static dramatization of political arguments, and instead solicits audience interpretation of the situations presented. Brechtian' theatre attempts to stimulate audiences' inner dialogues through alterations to existing conventions (e.g., estrangement) that model dialogic interaction and increase audience consciousness of its participation in the creation of meaning. Epic theatre presents working-class audiences – a constituency used by the powerful as a reservoir of passive compliance and politically valuable apathy – with social analysis through images and situations that provoke debate about the conflicts they encapsulate. Brechtian theatre models the dialectic thinking it seeks to promote in three important ways: (1) using direct address of the audience to comment on the dramatic action, (2) structuring narrative in the form of brief scenes familiar in agitprop, and (3) foregrounding theatrical mechanisms such as lighting to constitute them as a presence, both part of and separate from the representation. Brecht's alienation effects also serve to set up a perceptible dialogue between actor and character. Central to Brecht's project was the use of styles and conventions associated with working-class venues such as the music hall. In such venues, as Blackadder points out, audience response was not as constrained as in bourgeois theatre; interruptions in the form of comments or spontaneous applause were expected. Through the use of techniques such as a lighted auditorium, Brechtian theatre prevents the audience from assuming its mask of collective anonymity by acknowledging individual presence and identity. In his theoretical writings, Brecht urges the creators of political theatre to avoid traditional dramatic forms, with their narrative emphasis, stylistic unity, and emotional closure, in favour of theme-driven, episodic, open-ended, actor-focused productions laced with ironic humour, enlivened with music, and framed within a patently artificial historicized context (see Brecht, 1964).

Brecht's influence pervades contemporary theatre. Brechtian models, having made their way into the theatrical mainstream, no longer necessarily signal or define political theatre. Though Brecht's work remains an essential reference point for discussions of political theatre, contemporary practice questions its reliance on the intellectual at the expense of emotional and sensual experience, and the changed context created by what Fredric Jameson terms 'late, consumer, or multinational capitalism'

(Foster, 1985). Brecht's rejection of Aristotelian dramatic form does not necessarily convince contemporary dramatists; plays that evoke the 'pity and fear' of classical tragedy use strong emotion to counter indifference. At the same time, Brechtian ideas continue to provide the impetus for political theatre, especially through the use of frankly theatrical styles and the technique of juxtaposing representational and presentational moments. The important potential of Brechtian theatre remains its insistence on dialogue and its openness to active interpretation.

Antonin Artaud, whose theoretical work focused on articulating a 'theatre of cruelty', has made an impact on theatrical production and criticism that is often contrasted with that of Brecht. While Brecht emphasized theatre's potential to engage the audience in rational consideration of possible choices and outcomes, Artaud emphasized the non-rational elements of decontextualized images, inarticulate sounds, and strong, even painful, stimulation of the senses. By decentring and deconstructing language and by linking the conventions of the theatrical 'contract' to the operation of a hegemonic system of power, Artaud ushered in the theatre of the absurd and prefigured postmodernism. While Artaud's ideas find their purest expression in the postdramatic theatre described by Han-Thiess Lehmann, Artaudian tactics such as assaulting the senses with painful stimuli, deconstructing language, and decontextualizing familiar images play an important role in contemporary political theatre.

A number of contemporary writers and theatre practitioners have experimented with new ways of experiencing the power of theatre and using performance as an activist force in society. Among contemporary practitioners, the most influential in terms of expanding and developing the practice of political theatre has been the Brazilian-born Augusto Boal. His work, which has been international in scope, began as Theatre of the Oppressed, a type of street theatre that masqueraded as unrehearsed reality. It has more recently developed into Forum Theatre, a performative dialogue that constructs and addresses issues identified by participants. Forum Theatre eliminates the traditional dividing line between those who create the performance and those who receive it. Leaders engage participants as 'spect-actors', and support them in the process of articulating their own stories and strategies for change (see Boal, 1998). Rather than using written scripts, the participants create and revise performances that both bring political issues to visibility and create (and debate) interpretations of those issues. Productions based on Boal techniques, therefore, take the form of workshops rather than traditional theatre productions, and demand participation from all who attend the event. Unlike traditional theatre, which offers audience members only an illusion of community, Boal workshops demand that audience members interact with one another, form a temporary community, and determine which issues to bring to visibility within that community. Thus, they offer new potentials for understanding power through the formation of conscious, though transient, communities engaged in a political process. The impact of Boal's work is limited, however, by the fact that workshops select an even narrower segment of a given society than does traditional actor–spectator political theatre, because of the greater commitment involved.[10]

In England, the kind of involvement envisioned and advocated by Boal has been most evident in community theatre events. Community theatre is characterized by broad inclusion in a creative process based in a temporary or permanent community. Expanding the format of the Boal-type workshop, it encompasses a wider range of the populace than typically produce or attend traditional theatre. Its unique quality lies in the active creative work of non-professionals who in traditional theatre would be the audience, but who in this format have the opportunity and responsibility to take the stage. Community theatre productions have been created by towns and cities, schools, and prisons. Developed over a period of weeks or months by volunteers guided by theatre professionals, the production may commemorate a historical event, address a problem, or celebrate a milestone for the community. Productions may focus around a single narrative or image or assemble unrelated acts. They may employ music, puppetry, sculpture, and pageantry. Since its inception in the 1970s, community theatre has given rise to ongoing events, such as the annual Notting Hill Carnival, and ongoing companies, such as Banner and Welfare State International. Its primary theorist, Baz Kershaw, identifies the potential in community theatre for 'radical democratic empowerment' (1999: 80). Kershaw sees such productions as 'actively engaged in widening the bounds of political processes' and 'opening up new domains of political action', rather than indirectly modelling ideologies or reflecting political positions (1999: 84).

Contemporary political theatre has been crucially influenced by the emergence of postmodern theory in the late twentieth century. As represented by Beaudrillard, Derrida, and Foucault, postmodern theory negates concepts of identity, agency, and community that have been central to political theatre – and, for that matter, to political action in general. In questioning oppositions between reality and illusion, public and private, nature and culture, postmodern thought has tended to deny theatre a distinctive function and politics any prospect of change.

Irony remains as the primary function of art, and in this spirit Gilles Deleuze calls for a 'theater of nonrepresentation' (1997: 241). Deleuze describes the political aspect of non-representative theatre in this way:

> You begin by subtracting, deducting, everything that would consti-
> tute an element of power, in language and in gestures, in the repre-
> sentation and in the represented. You cannot even say it is a negative
> operation because it already enlists and releases positive processes.
> You will then deduct or amputate history because History is the
> temporal marker of Power. You will subtract structure because it is
> the synchronic marker, the totality of relations among invariants. You
> will subtract constants, the stable or stabilized elements ... You will
> amputate the text because the text is like the domination of language
> over speech and still attests to invariance or homogeneity. You deduct
> dialogue because it transmits elements of power into speech ... you
> deduct even diction and action ... Everything remains, but under a
> new light with new sounds and new gestures (245).

Hans-Thiess Lehmann, in his 1999 study of European theatre at the turn of the millennium, offers a view that is both comprehensive and analytic of what he terms 'postdramatic theatre' (2006). Tracing its development out of the well-known theatrical currents of the twentieth century, Lehmann describes postdramatic theatre as production of 'inner and outer states' (68) through performance and the technical elements of theatre, but without the dominance of a narrative. A text or narrative may be part of the production, but will not take precedence, instead existing in a non-hierarchical relationship to other elements. Postdramatic theatre uses aspects of ceremony but resists synthesis. It substitutes narration for narrative and relies on the potential of simultaneity that is inherent in theatre semiotics. Postdramatic theatre may 'play with the density of signs' (89) by minimizing the density to stimulate the audience's imagination and creative collaboration or offering a proliferation of signs to challenge the limits of materiality and create disorientation through sensory overload. The sights and sounds of postdramatic theatre privilege purely sensory over linguistic experience and are based in a materiality that, in its 'nonmimetic but formal structure', is essentially 'concrete' (98). The actor's body is taken as 'the central theatrical sign' (95). In postdramatic theatre the body 'refuses to serve signification' (*ibid.*) but may perform 'irruptions of the real' (99) that negate aesthetic unity. Lehmann gives examples of the postdramatic aesthetic in the work of Robert Wilson, Heiner Müller, Peter Handke,

Jan Lauwers, and others. Explicitly addressing the political potential of postdramatic theatre, Lehmann acknowledges that the nature of the form excludes consideration of issues and problems, but suggests that its political impact lies in engagement with 'the forms of perception' (184) to 'make visible the broken thread between personal perception and experience' (186) and thereby potentiate an 'aesthetics of responsibility' (184).

While Lehmann's analysis of postdramatic theatre offers useful ways of understanding late- twentieth-century theatre, there is no question that postmodern theory and its theatrical offshoots have played a part in delegitimizing socially activist theatre and inhibiting recent development of issue-based drama. Social theorists such as Linda Nicholson, Chantal Mouffe, and Iris Marion Young have expressed dismay at the disabling of political activism through the critique of identity and agency, which seems to play into the hands of politicians like Margaret Thatcher, who famously taunted social activists by declaring, 'There is no such thing as society' (*Women's Own* magazine, 3 October 1987). Nicholson, Mouffe, Young, and others have critiqued the interaction of poststructuralism with social theory, noting the impossibility of theorizing positive social change within poststructuralist theory. Nicholson observes,

> Poststructuralism, in particular, but postmodernism also, became significantly associated with a critical mode of analyzing texts. At times, the social was collapsed into the textual … As important as deconstruction was to politicizing language and knowledge, this 'textualizing' turn of the postmodern meant that many of the issues that have been pivotal to social theorists were neglected (8).

Nicholson concludes that, despite its value in the 'substitution of a politics of difference for a millennial liberationalist politic', poststructuralism has 'turn[ed] us away from the social' (8–9). Nicholson and others have gone on to formulate paradigms of political action that use positionality rather than identity as the starting point for social action.

Postmodernism and the critiques it has generated are both reflected in contemporary political theatre. Productions impelled by postmodern thought evoke states of psychic disintegration, materialist domination, and social paralysis by means of a non-mimetic theatricality that, in the words of Timothy Murray, constitutes 'an unbalanced, nonrepresentational force that undermines the coherence of the subject through its compelling machineries' (1997: 3). Postmodernist thought is deliberately explored in the plays of Martin Crimp, Sarah Kane, Mark Ravenhill, and Patrick Marber. Jon Erickson identifies postmodern theatre particularly with performance art monologues that, in their 'radical distrust of language', demonstrate a 'hermeneutics of paranoia' (2003: 176) and hold only the limited political potential of 'resistance to, or evasion of, hegemonic subject production' (173).

In a study of theatre, a live collaborative art that necessarily involves the convergence of multiple perspectives, different types of work, and variously influenced audiences, one must be cautious in the use of theory. Australian director Julian Meyrick, in a 2003 essay, critiques the 'estrangement from actual practice' (231) in the writings of contemporary theorists on theatre. Starting from actual practice in contemporary British theatre reveals a distrust of ideologies and resistance to established formulae. It shows conscious but eclectic juxtapositions and combinations of viewpoints, techniques, and styles. It also points to influences from beyond theatre, including current styles and performance conventions in popular music and film. Contemporary theatre practice does, however, seem strongly and perhaps inevitably influenced by the theatre practice and ideals of Peter Brook, as it attempts to stage what has been invisible and involve audiences in experiences that create the sense of immediacy and urgency.

A constellation of material factors supports the creation of political theatre in Britain. An extensive infrastructure sustains theatre production, and many resources are devoted to the generation of new plays. Non-commercial London theatres like the Royal Court Theatre and the Theatre Royal Stratford East, as well as regional theatres like the Liverpool Everyman or Manchester's Royal Exchange, maintain long-established programmes to develop new plays. Smaller and more narrowly focused theatres such as the Donmar Warehouse, the Hackney Empire, the Soho, the Bush, the Hampstead, the Almeida, the Tricycle, the Riverside Studios, and the Young Vic are also known for producing new plays, including those with political themes. In recent years, Britain's flagship theatre, the National, has devoted many of its considerable resources to new plays. The London infrastructure also, of course, includes the commercial theatres of the West End and the numerous Fringe theatres in rooms above pubs and other converted spaces – the former used for the transfer of particularly successful plays first produced in non-commercial theatres and the latter providing performance space for smaller touring productions. Regional theatres, in addition to their own new play initiatives, contribute further to the development of political drama through co-productions with London theatres.[11]

Britain's touring companies constitute an essential element in the creation of political theatre. Operating without a permanent performance venue, touring companies take productions to performance spaces throughout a particular region, or on a national, and sometimes international, tour. These companies may also co-produce some plays with building-based theatres, thereby lessening the financial risk of presenting new plays. Their varied focus and wide geographical distribution expand the range of issues, forms, and performance styles encompassed by contemporary political theatre. The most visible of the companies dedicated to new writing – Out of Joint, Paines Plough, Hull Truck – as well as a host of smaller ones, generate new work by commissioning playwrights of interest or, less frequently, collaborative development by a creative team.

The extent and range of theatre activity in Britain could not occur without substantial public funding. Two organizations, the Arts Council and the National Lottery, allocate funding for the arts. There are actually three separately operating arts councils in the United Kingdom – Arts Council England, Arts Council Wales, and the Scottish Arts Council – but the largest by far is Arts Council England, which currently administers £369 million per year received as grants-in-aid from the Department for Culture, Media, and Sport. Mandated to keep itself at 'arm's length' from the government in order to assure artistic freedom, the Arts Council operates on the premise that 'the arts have power to transform lives, communities, and opportunities for people throughout the country' (www.artscouncil.org.uk). Arts Council England supports theatre through several types of funding for organizations and individuals. Currently it contributes to the core costs of nearly 300 organizations, with the yearly funding amount ranging from £16 thousand to over £17 million. Regularly funded organizations include large and small building-based theatres in cities, touring companies based throughout England, multi-purpose arts centres in towns, audience development organizations and booking agencies in rural areas, and organizations that hold and promote festivals. The National Lottery, a non-tax means of funding the arts and national heritage projects, has contributed £441 million to theatre remodelling projects since its inception in 1995.

Specific organizations and individuals played important roles in the resurgence of political theatre. The Royal Court Theatre, in its role as the most influential new writer's theatre in Britain, premiered many of the plays that constitute the revival of political theatre, under the artistic direction of Stephen Daldry in the 1990s and Ian Rickson after 2000. Out of Joint, the company directed by Max Stafford-Clark, has developed and produced a rich stream of political work. The *Transformation Season* at the National Theatre, in 2002, spearheaded and personally financed by Trevor Nunn, the National's artistic director at that time, brought political theatre to this important and visible venue. Under the current artistic direction of Nicholas Hytner, the National has produced and co-produced some of the most significant recent new political theatre. Tricycle Theatre, with its director Nicolas Kent, introduced the tribunal and verbatim plays that have become a central element of contemporary political theatre.

As Raymond Williams suggests in *The Long Revolution* (1961), new formal structures in the dramatic arts arise in response to developments in society. Theatre in post-Thatcher Britain has given rise to new strategies of communication, along with long-established forms, to confront current political realities. A strategy that generated much attention and excitement in the late 1990s was the use of extreme or unusual forms of violence. Plays that assault the sensibilities of the audience with graphically presented acts of violence seem to use shock as a unifying reference point that provides entry into a shared constructed reality akin to ritual. While these plays rely primarily on what Patterson has called a reflectionist (in contrast to interventionist) strategy, they intensify the reflection to the point that its very intensity creates a type of intervention. The prevalence of characters and situations that represent the extremes, rather than the typical dimensions, of society – in effect, making the exception the norm – along with the literalness of representation, has given rise to a style that might be considered a return to naturalism except for the inclusion of markers that make tangible the process or devices of construction. A second notable strategy that has appeared since 1995 uses theatre to present documentary-type reports on controversial events. These plays make an explicit attempt to counter the influence of mass media, especially television, in the selection and viewpoint of material. Styles from the past used alone or in combination include epic theatre, social realism, absurdism, and satire.

Individual voices predominate in the contemporary political theatre. Despite the singular mode of expression, however, common themes point to the shared political and theatrical system. Themes often seem to hearken back to the decade of the angry young men, in their evocation of frustration and anger, but the current generation also expresses loss – loss of the strong sense of purpose associated with social movements, and loss of the exhilarating experience of collective passion generated, at least for a time, by such movements. While most contemporary work rejects idealism, contemporary dramatists leave open the possibility of change, locating this possibility in the centrality of dialogue. The plays open up dialogues among family members, friends, workmates, people brought together by a particular issue. They imagine dialogues that have never happened, and perhaps will not. They create a rich dialogic relationship with the audience.

An outpouring of significant plays by younger dramatists is one of the most important aspects of contemporary British theatre. Works by new playwrights use a wide range of styles, but share a determination to present an unvarnished look at British society. These works especially pay attention to the elements of society usually hidden from public view – unemployed young people, mental patients, drug addicts, homeless vagrants – absent in commercial theatre and even, it can be argued, glossed over by the alternative companies of the 1970s and 1980s, which tended to emphasize idealistic appeals for change rather than the depth of particular problems. The current generation of young playwrights, having witnessed the recent failures of both the left and the right, focuses on entrenched problems not adequately addressed by political parties or processes. Their anger contains a great deal more bitterness than optimism. Their view of the pervasiveness of social issues connects the most everyday and intimate acts of personal life to structures of power in British society. Though the majority of new playwrights are male, and the themes of their plays decidedly masculine, young women have made an impact individually. Feminism, however, does not constitute a central aspect of their plays, and several of the new generation have made it clear that they do not want to be categorized as women.

Established playwrights, as well as the new and unknown, have contributed to recent political theatre. Many older playwrights have responded to issues introduced by the post-Thatcher generation, creating a dialogue between the generations that examines history, responsibility, and the potential of action. Well-known playwrights have embraced broad themes relating to history and structures of power. Plays addressing both internal issues and problems involving Britain's exercise of power within the global environment contribute to a broad-ranging critique of Britain's political choices. These plays, too, offer individual perspectives, and often place global issues in an almost autobiographical framework.

When the collective voice is heard in contemporary British theatres, it most often occurs in work that expresses and interprets the social position and culture of a particular ethnic or regional group within Britain. Of particular note are the West Indian and South Asian immigrant communities that have established a significant presence in Britain during the past 40 years. These communities have not only brought with them cultural institutions and practices but also created or participated in new institutions through which they negotiate their engagement with British society and its dominant culture. Plays that employ specific cultural contexts necessarily reference common experiences and perspectives of that cultural group. By making visible specific cultures, these plays also claim inclusion in the broader public life of the nation. Works representing ethnic identity and communities project a more optimistic mood than do other contemporary works, as they mirror movement, progress, and gain, as well as loss.

As this study will show, theatre is making an impact on current understandings of society and politics in contemporary Britain. In spite of the suppression of activism during the Thatcher era and the post-Thatcher mood of political ennui, the upsurge in political playwriting has brought in a new generation of playwrights and re-energized an older generation to create fresh sociopolitical critiques. The most powerful quality in contemporary political drama is the immediacy of its impact. Physical and emotional distance shrinks, as both actors and audience seem to seek an unmediated connection. Participants on both sides of the stage divide speak of a sense of risk and danger, but this risk factor has proved to be a powerful attractant.

Given the renewed strength of theatre culture in Britain at the same time that political life is characterized by public disengagement and detachment, the theatre's capacity to provide a space of freedom becomes even more crucial to the possibility of addressing particular issues or broader ones of social change. The exchange of energy between performers and audience in the live production in itself constitutes an action, rather than a passive state of being. When a production addresses contemporary politics, even to emphasize the intractability of problems, it negates an attitude of hopeless cynicism in relation to those problems. The commitment to action inherent in performing or attending a theatre production will tend to promote, impel, or even extend to direct forms of political participation. The current situation demands both new political strategies and new artistic forms. Effective political theatre today must acknowledge the current mood without succumbing to it. It must call upon the social, cultural, and political energy of those who have been excluded from or marginalized in definitions of the nation. A striking array of playwrights, directors, and companies have taken on this challenge. As it engages with contemporary politics, their work is reinventing political theatre.

2
Generational Politics: The In-yer-face Plays

In the mid-1990s, a new impulse shook up British theatre and altered its direction. Young and unknown playwrights working in small-scale venues associated with new writing confronted audiences with amoral characters, shocking scenes of violence, and crudely explicit language. Sarah Kane, who has become an iconic figure of the post-Thatcher generation, was the first of these young playwrights to gain widespread attention, though the initial public reaction was almost unanimously negative. When Kane's play *Blasted* was performed at the Royal Court Theatre Upstairs in 1995, critics catalogued its surreal but explicit acts of violence, noted the inexperience of the 23-year-old playwright, and condemned the play as a juvenile attempt to shock audiences. The play turned into a *cause célèbre* with extraordinary media attention, defences of the play by prominent playwrights, sold-out performances, and a new interest in theatre by younger audiences. The turbulent reception of Kane's debut play focused public attention on the presence of young dramatists and the urgency of their voices. The apparent drama of youthful anger breaking through a miasma of tradition may have owed more to the received memory of the angry young men of the late 1950s than it did to the actual unfolding of events in the 1990s. However, it did create the condition for recognizing writers born around 1970 as a generational cohort, and it did encourage those writers to express their desires, fears, and anger in terms that can be identified with a particular historical moment.

Within five years, Kane's plays were being thoughtfully reconsidered and remounted,[1] and a large group of young playwrights were having their plays produced in the subsidized theatres, with occasional transfers to the West End. Kane's plays were considered in relation to other young playwrights. Some plays of the early 1990s, such as Philip Ridley's

Pitchfork Disney (1991) and Anthony Neilson's *Penetrator* (1993), were now seen as prototypes for the new drama. Aleks Sierz gave the collective style the descriptive term 'in-yer-face theatre' (2000). The young writers who form the core of this generational cohort have also been called 'new brutalists', 'new Jacobeans', 'theatre of urban ennui' (Sierz, 2002), the 'new nihilists' (Urban, 2004), and the 'Britpack' (Saunders, 2002: 5). These terms have been the subject of continual challenge and revision, even by the critics who coin or use them, but Sierz's term 'in yer face' has become the most commonly used. It evokes the immediacy and the confrontational quality that breaks boundaries, arousing a sense of uncertainty and risk. Many, if not most, of the in-yer-face plays – even those written by women – employ a pugnacious tone that has been referred to as 'laddish'; this characteristic evokes Margaret Thatcher's culture of toughness and confrontational style. Finally, the in-yer-face plays seem motivated by a spirit of impatience – a demand that they be seen and heard now – that is quintessentially youthful.

The playwrights included in descriptions of in-yer-face theatre do not, of course, speak as one; nor do they acknowledge a common cause or profess loyalty to any ideology. The individualistic standpoint forms a hallmark of their work, and it too evokes the tone of the Thatcher era. The in-yer-face playwrights do, however, evince awareness of their generational commonality and often acknowledge one another in their public statements. When Mark Ravenhill writes of his work being precipitated by the murder of James Bulger (2004), and Gregory Burke admits to wearing a 'Prada sport jacket and a pair of trainers made in Vietnam' while participating in an anti-globalization discussion (2003), they speak from a base of generational experience and issues. The in-yer-face playwrights also make it clear, through descriptions of characters by age, that their work focuses on members of their generation. The plays participate in a common desire to characterize the specific culture of 1990s Britain and people coming into adulthood during that time period. The plays highlighted in this chapter are those that most consciously explore the political phenomenon of the first post-Thatcher generation.

The political significance of the in-yer-face plays has been called into question because the anti-rhetorical form they often take has not generally been associated with British political drama. Benedict Nightingale, who refers to the playwrights born around 1970 as 'Mrs Thatcher's disorientated children', sums up the type of plays they write as 'snapshots' of gang members, criminals, and misfits that provide entertainment but lack coherence (1998: 20). Nightingale's comment, however, signals the unique political dimension of this generation's plays: they are by and about 'Thatcher's children'. Post-Thatcher playwrights initiated a dialogue about themselves in relation to the previous living generations in Britain – and especially in response to the concerns of their parents' generation. Their plays have not only quickly taken up a space in the collective consciousness but also stimulated an outpouring of additional plays that address generational issues. The dialogue between generations has proved to be the most salient feature of post-Thatcher drama. In this period of transition, the generational dialogue debates the meaning of the past, assumptions about the future, and the possibilities for societal change.

What does it mean to be a child of the Thatcher era – someone who grew up during the time when the Conservative leadership was remaking British society in accordance with its creed of personal responsibility, centralized government, and market primacy? Sierz offers a compelling synopsis of their socio-political experience:

> One way of understanding the point of view of the young writer is to... imagine being born in 1970. You're nine years old when Margaret Thatcher comes to power; for the next eighteen years—just as you're growing up intellectually and emotionally—the only people you see in power in Britain are Tories. Nothing changes; politics stagnate. Then, sometime in the late eighties, you discover Ecstasy and dance culture. Sexually, you're less hung up about the differences between gays and straights than your older brothers and sisters. You also realize that if you want to protest, or make music, shoot a film or put on an exhibition, you have to do it yourself. In 1989, the Berlin Wall falls, and the old ideological certainties disappear into the dustbin of history. And you're still not twenty. In the nineties, media images of Iraq, Bosnia and Rwanda haunt your mind. Political idealism—you remember Tiananmen Square and know people who are roads protesters—is mixed with cynicism—your friends don't vote and you think all politicians are corrupt (2002: 237).

The formative years of this cohort saw dramatic changes in the nation and the world. Important events include the economic crises of the 1970s, the cuts to public-sector institutions during the Thatcher years, the decline of trade unions, massive job losses in manufacturing and mining, and London's rise as an enclave of wealth. Traumatic events that played a part in defining the post-Thatcher generation (see Edmunds and Turner, 2002), include the AIDS epidemic, a savage civil war in the Balkans, the 1993 murder of 2-year-old James Bulger by a pair of 10-year-olds in Liverpool, and the racially motivated murder of 18-year-old Stephen Lawrence in London the same year.

In the post-Thatcher period, conditions have militated against the sense of personal security and power that provide a foundation for broad-based political involvement. The widening gap between rich and poor, a central trend in post-Thatcher Britain, propels other differences – in educational attainment, age of marriage, and rate of employment. An underclass is becoming entrenched. For those who do have education and employment, higher average incomes have brought not only better diet and improved housing, but also major increases in personal debt and bankruptcy. Even for middle-class young people, the high level of economic insecurity has created a climate of disempowerment.

Recent surveys of societal trends show declines in family resources, social support, and social involvement.[2] The in-yer-face plays of the mid-1990s reflect the specific historical forces of the Thatcher era and its aftermath, and they point to the way in which this era's events and trends shaped their individual and collective political subjectivities. Reflecting the plays' roots in social realism, the settings are generally ordinary places with a sense of familiarity if not comfort, and a tension between confinement and freedom. Dramatic styles draw upon Artaud's Theatre of Cruelty, Harold Pinter's theatre of menace, and Grotowski's Poor Theatre, as well as the work of contemporary film directors such as Quentin Tarantino. Dramatic structure often subverts narrative to negate history. Action depicts an unending and seemingly unendurable present. Characterization focuses on the writers' generational cohort and often presents a group of young people rather than a single individual as representative of that cohort. Despite their stylistic debts to earlier eras of theatre, the in-yer-face plays do not express a sense of continuity; rather, they signify a break with the immediate past. They constitute a rejection of Thatcherite policies and an equally strong refusal to return to the pre-Thatcher social agenda of the 1970s. Their break with the past does not imply future directions, but instead focuses on a frank and sometimes frightening portrayal of their generation as a political force.

Sarah Kane's visions of catastrophe

Sarah Kane's plays challenge audiences with their boldly experimental theatrics, neo-mythical form, and unrelenting focus on physical and psychic pain. In her first three plays, Kane employs structures of confinement to explore the political landscape of the post-Thatcher generation. The 'expensive hotel room' in which the action of *Blasted* occurs creates a metaphorical space for her generation's coming of age. While locating the hotel room in Leeds, Kane strips it of specificity by noting that the room could be found in any city. Fresh flowers and champagne on ice signal an impending celebration, but the entrance of the two primary characters, a 45-year-old man who acts as though he is being hunted and a 21-year-old woman who seems younger than her age, jars the mood of anticipation. The rough appearance and speech of Cate and Ian places them at odds with the luxurious hotel room. Ian nevertheless strides in with laddish bravado, proclaiming, 'I've shat in better places than this' (3).

The hotel room, a shelter that is not a home, provides the conceptual framework for understanding territorial power. Ian, whose power derives from the combination of knowledge and force, controls access, refers to the hotel staff in demeaning racial terms, and speaks insultingly to Cate. Cate's limited power consists entirely in her physical separateness and difference from Ian; thus their power differential takes the form of gender oppression. The bed, which dominates the space, becomes the locus of a disturbing conflict characterized by overtones of paedophilia or incest, and frequent eruptions of violence. Cate, propelled into fits of stuttering and fainting by stress, seeks in the bed the comfort associated with a parent's reassurance after a bad dream. To Ian, who lives in a continual nightmare of real or imagined threat, the bed is an area in which he can assert his will securely. As implied in the dialogue, Ian rapes Cate vaginally and anally during their night together, extending his territorial power to her body. In the morning, Cate reasserts her separateness by ripping Ian's jacket, clamping her teeth on his penis during oral sex, and locking herself in the bathroom.

The sudden invasion of a rifle-wielding soldier negates Ian's control or illusion of control. The soldier takes Ian prisoner and forces open the bathroom door, revealing that Cate has escaped through a window. Then a bomb blast rips open the room, destroying its function as a refuge and rupturing the play's conceptual boundaries and realistic conventions. The destruction of the room's boundaries exposes Ian to powerlessness. Propelled by the desire for vengeance, the soldier rapes and mutilates Ian, in action that mimics both Ian's previous violence and the violence of the war that has suddenly engulfed them. The catastrophe conversely signals a kind of liberation for Cate, enabling her to move beyond victimization. When she returns she appears to have gained autonomy and discovered a potential for love, despite the violence. She carries a fragile emblem of hope for the future, a baby she is trying to save. Cate seeks food for the baby, but this room has never contained food. When the baby dies, she buries it in the floor beside the bed, fashioning a makeshift cross as a marker.

The invasion of the room destroys Ian's individual identity. The soldier treats him as a faceless enemy responsible for the atrocities enacted by his side in the war. The soldier's consummate act of vengeance, sucking out and eating Ian's eyeballs, removes the last vestige of Ian's power as it gives tangible form to his terrible inability to genuinely see others' suffering despite his news reporter's job of observing and recording suffering. Ian ends the play helpless and without hope – his retreat smashed, his gun empty, his body maimed, and his newspapers used to clean up after he soils himself. He seeks a grave in the floor beside the baby and dies, but his death does not remove him from the present. In an act of savage futility, he eats the baby's corpse and thus destroys even the symbolic remains of future life. The quality of violence that engulfs the room has changed from domestic and realistic to mythical and nightmarish. Nevertheless, Cate returns once more, bringing food that she shares with Ian in the final moments of the play.

Kane's Cate comes of age in an environment that mirrors the glamorous settings of Hollywood's romantic fictions but functions as a trap. The ostensible protection of the possessive and violent Ian turns into humiliation and injury. Cate attempts separation and escape, but her attempt is overwhelmed by the intrusion of a war that outdoes Ian in cruelty. As Ian is rendered helpless, and the soldier kills himself, the visible symbols of power disintegrate. The resulting anarchy, though full of dangers, gives Cate the opportunity to nurture the embodiment of the future. Even when the baby dies, Cate tries to protect and respect her through burial and prayers for her future, 'just in case' (58). In the final moments of the play Cate frees herself from the past by feeding the now victimized Ian and receiving his spoken expression of gratitude – the only moment in the play when he respectfully acknowledges another person. At the end, she has neither past nor future, and oppositions meet in her image and actions. Her actions suggest an other-worldly spirituality, but the blood dripping down her legs and her meal of sausage and bread emphasizes physical need and implies that she obtained the food in exchange for sex. No longer a child, but not fully adult, Cate sits quietly sucking her thumb, providing primitive comfort to herself in a damaged room that offers no definition, protection, or certainty, but that seems to be her only home.

As a representative of the post-Thatcher generation, Cate stands between past and future without really knowing either. Though she seems introspective, Cate says little, speaks hesitantly, and falls into unconsciousness. Unable to discuss ideas, Cate reveals herself through her actions. She joins aspects of masculinity and femininity, presenting a boyish appearance but enacting the stereotyped feminine roles of mother, prostitute, and angel of mercy. She seeks in the previous generation comfort and protection, but its representative Ian dominates and harms her. Her actions show the capacity for compassion, nurture, resistance, and survival. A commitment to living is shown in her refusal to give Ian the means to end his life and by eating the meat available to her in spite of her vegetarian principles. Though she survives much, her future remains unknown at the end of the play, when she is the only living character on the stage. Cate has failed to save the future represented by the baby, but her hopes for it are so strong that she offers prayers for its life even while burying the corpse. The cruelly bleak eternal present she inhabits offers no apparent potential for connection and no means by which to build a life beyond mere survival. Cate's fragile physical presence serves as the sole indicator of hope in a world that has battered her with personal and societal violence and exposed her to the direst threats.

While Cate's living presence signals hope for the survival of the love and generosity she has demonstrated, in spite of the violence surrounding her, it offers very little hope for the mobilization of those moral qualities to create a better society. Cate herself is childlike and incapable of dealing with aggression or threats. She has found no ally in Ian or protection from the room amid the chaotic violence of war. Her love for the baby does not give her the means to provide for its needs and keep it alive. The hostilities that engulf them mete out a grim and primitive form of justice to Ian for the wrongs he has committed, but offer no form of support for Cate's moral choices. The small hope for change lies in the personal transformation of Ian. As he suffers and dies, he is able to receive and comprehend Cate's kindness, and his last words are words of thanks.

Phaedra's Love, first performed at London's Gate Theatre in 1996, moves decisively away from social realism, taking an ancient myth as its point of departure and employing a style of absurdist comedy. The play's setting, a darkened room littered with dirty clothes, suggests, in Graham Saunders's view, the imprisonment of Hippolytus by 'gross appetites' that arise from his attempt to fill a vast interior void (74). A pampered narcissist mired in permanent adolescence, Hippolytus watches television, plays with electronic toys, eats junk food and sweets, and masturbates into a sock. He isolates himself from all except women who perform sexual services. Despite his passivity, sloth, obesity, and dirtiness, 'everyone loves him', including his stepmother Phaedra; but as his stepsister Strophe observes, 'he despises them for it' (72).

Hippolytus exemplifies one of the extremes Saunders has identified in Kane's plays. The other characters provide a series of contrasts. The constant absence and public activities of his father Theseus contrast with Hippolytus's immobility and solitude. Phaedra's passion presents the antithesis of Hippolytus's indifference. She raves about the desire 'burning' her (and moves inexorably towards the fire of her funeral pyre), while Hippolytus insists, 'No one burns me, no one fucking touches me' (83). Strophe, Phaedra's daughter by a previous marriage – and Kane's addition to the characters of the myth – diverges from Hippolytus in her sense of responsibility and concern for others. Strophe, though variously disregarded and exploited by members of the royal family, maintains a steadfast loyalty to it. She serves as confidante to her mother and to Hippolytus, doing her best to protect them from the desires that lead them towards destruction, even while lacking the power to do so.

This royal family defines itself through power rather than by blood relationship or love. Both the idolized Hippolytus and the ignored Strophe are creatures of a system that distributes power unequally. He is a prince, with contemporary references to the British royal family and to entertainment celebrities. In demand because 'everyone wants a royal cock' (74), he indulges in sex indiscriminately but shares himself with no one. Knowing that showing emotion would diminish the extraordinary power he maintains through withholding, he does not reveal his feelings, even when he ejaculates during sex. Strophe, the only one in this artificial construct of monstrous egoists who shows concern for others, has no power, status, or distinction.

The blackly comic action of the play focuses on the interaction of these extremes and the resulting destruction of the family. Phaedra visits Hippolytus on his birthday to present him with a gift. Ignoring his repulsive behaviour, she performs fellatio on him and affirms her love as he insults and taunts her. After Phaedra has gone, Strophe brings Hippolytus the news that Phaedra has hanged herself, leaving a suicide note accusing him of rape. Hippolytus understands Phaedra's fabricated accusation as the promised gift, and it jolts him out of his indifference and paralysis. Rather than hide, as Strophe urges, he leaves his room and turns himself in. Imprisoned, Hippolytus is visited by a priest, with whom he engages in spirited debate about God, sin, and honesty. Hippolytus insists that if he were God, he would condemn only those who 'try to cover their arse' (96). The priest acknowledges in Hippolytus 'a kind of purity' (97) before he kneels and, like Phaedra, performs fellatio on him. In the final scene, Hippolytus is brought out for execution through a crowd that has gathered with picnics and barbecues. Strophe and Theseus, disguised, mix with the crowd. Strophe tries to defend Hippolytus from attacks. Theseus rapes and kills Strophe without recognizing her. He sees and is acknowledged by Hippolytus but delivers him to the mob, which cuts off the 'royal cock' and tosses it on a barbecue grill, among screaming children, and finally to a dog. In the final moments, Theseus realizes what he has done to Strophe and cuts his own throat. As vultures descend upon the bodies, a still-conscious Hippolytus exults.

The play's two young characters offer a novel perspective on idealism. Both Hippolytus and Strophe die for ideals. Hippolytus takes action when Phaedra's incontestable accusation of rape confers upon him an identity he can accept – an 'exciting' identity in which he finds 'life at last' (86, 90). Of course, this identity is false, as he did not rape Phaedra, but from the moment he assumes it, Hippolytus adheres to it with fanatical consistency. Strophe, as she has promised to do, dies for the ideal of the family, with a loyalty that is unswerving but arbitrary, since she is not related by blood or reciprocated affection. As Hippolytus observes, Strophe is 'the one person in this family who has no claim to its history' (88). Their deaths reflect the differential power of their positions in life. In being torn apart by the masses, Hippolytus shares himself with them, becomes their victim, and ensures himself a place in their mythology. His choice of this death makes him the author of his own story – a story that will forever mask his real identity as 'a fat boy who fucks' (88). Thus, Hippolytus dies in a moment of supreme realization of his personal power. Strophe, by contrast, dies anonymously and by chance, with no survivor to tell her story. Her existence, which was never central, is merely erased, ensuring that she will not be part of the family's story. This family has been both created and destroyed through the power of public adulation. The family's actual destruction enables its power to endure, as it is transmuted into myth.

Kane's revision of the myth, however, introduces an element of subversion with the character of Strophe. Strophe, which means literally to turn, is, in a world of egos gone mad, the only one who listens to others. Her listening is the device through which Kane deconstructs the traditional myth of Phaedra and Hippolytus, along with the more pedestrian myth of the family. Strophe's ability to listen gives the audience access to an alternative view. What Strophe hears enables the audience to turn away from the fascination with celebrity and royalty and from prevailing myths of the family. In Kane's revision, Strophe is the almost overlooked representative of her generation who sees the falsity of all available ideals and sacrifices not only her life but also her identity to discredit the most powerful of these ideals.

Kane's first two plays offer models of political subjectivity that point to the possibility of meaningful action in situations of extreme powerlessness. In both plays Kane places a male/female pair at the centre of the action. While the barriers to action in the first play arise from territorialism and violence and in the second from the disintegration of the family, in both plays the actions of the central woman bring about some form of change. In spite of their lack of power, the young women move from being victimized to asserting moral authority. These plays, then, offer signs of hope, but those signs are almost lost in the overwhelming sensory effect of the elements of cruelty and suffering. The immersion in darkness became even more intense in Kane's third play.

Cleansed, which premiered at the Royal Court in 1998, locates its action in an authoritarian institution that harshly enforces narrow definitions of normality. This institution is initially identified as a university surrounded by a boundary fence, and locations within it are named 'the university sanatorium' and 'sports hall' (112, 116) in Kane's text. These identifiers, however, prove unstable, as the locations expressionistically transform into a hospital, a torture chamber, and a peepshow booth. The paternal Tinker (mischievously named for the *Daily Mail* critic who led the charge in condemning *Blasted*) shows corresponding instability. In the first scene he states that he is 'a dealer, not a doctor' (107), but is addressed as doctor, though he harms rather than heals. Tinker rules the institution and controls the students or patients or inmates who are confined there with an unremitting cruelty that is both intimate and unspeakable. Tinker adapts his tortures to the desires or conditions of the five characters confined within the institution. Grace and Graham, a sister–brother pair, exemplify an incestuous desire so acute that it seeks the annihilation of the self in the other. Carl and Rod are homosexual lovers. Nineteen-year-old Robin lacks the ability to express his yearnings in writing or speech.

Desire as experienced by the confined characters in *Cleansed* negates stable identity and bounded ego. They exchange clothing, identities, and even body parts in a pageant of transformation that evokes dreaming and the subconscious mind. As the 'doctor' – and the only character with a consistent persona – Tinker learns the secret desires of each person under his control and performs on them diabolically crafted treatments that enact their wishes but bring even more pain. In the first scene he responds to Graham's plea for a stronger dose of heroin by injecting him in the eyeball with a lethal amount. In response to Carl's desire to prove his love for Rod, he subjects him to painful beatings and impales him through the anus, offering to substitute Rod as the victim if Carl gives the word. When Carl fails this test, he feels the need to atone, and Tinker serves him by progressively dismembering him each time he expresses his love for Rod. Tinker tests Rod, too, killing him instantly when he passes the test by accepting death in order to spare his lover. When Grace wishes that her body were like Graham's, Tinker performs crude sex-change surgery on her, using Carl's severed penis. When Robin tries to express his inarticulate love for Grace by bringing her chocolates, Tinker forces him to eat them until he is sickened, distressed, and humiliated.

As a counterpoint to the tortures, each of the characters achieves, at some point in the course of the play, a moment of intense realization of love. These moments use theatre as poetry, evoking an inexpressible tenderness with vivid images, as when Grace and Graham reunite. He does a dance of love for her, which she imitates until they move in perfect unison. She initiates lovemaking, and their pleasure blooms in the form of a sunflower that bursts through the floor. Even Tinker experiences a moment of joy. He has compulsively viewed a woman dancing through a peepshow window, disturbed by the power exerted over him by a spectacle that he ostensibly controls. When the woman who has been confined to the other side of the peepshow window comes through this artificial frame and offers love, Tinker, though initially alarmed, allows himself to be drawn into her embraces. He enquires her name, which she gives as 'Grace', and finally expresses his love for her. This transformed Grace abandons the limiting but safe enclosure of the peepshow, which keeps her at a distance from Tinker, to enter his world of intimate cruelty. It is her action, full of danger and uncertainty, which changes Tinker into a human capable of giving love.

Cleansed places suffering and love as constants in a labyrinthine social landscape based on oppositions such as love and hatred, or pleasure and torture, but torn by the continual and confusing mixing and melding of these oppositions. The need to escape from suffering strongly motivates action, but love and suffering cannot be separated, and the only escape is to develop lack of feeling. Feeling, however, permits the moments of exultant transcendence when love is expressed, and thus the desire to express love proves more powerful than the need to escape pain. This play, in common with Kane's previous two works, suggests that extreme pain may transform the sufferer and that powerlessness may open him or her to a supreme experience of love. The world and its conditions as Kane creates them are at once personal and generational. Grace, a young woman uncertain of identity but filled with desire, begins to experience love and struggles with unbearable pain. She overcomes victimization and expresses the full dimensions of her unique love despite confinement within an institution that constrains the young within arbitrary boundaries and uses torture to enforce its norms.

Cleansed offers a stronger note of hope than Kane's previous plays. Again a young woman serves as the agent of change. In a moment of significant decision Grace breaks through the frame of objectification and confronts the authority figure Tinker with recognition of his humanness. She asks for recognition of her own humanness, and in this moment when control is taken from him Tinker gives this recognition and expresses love. Though Tinker's transformation does not dissolve the boundaries of the institution, it fosters a regeneration of life for the maimed souls within it. Through endurance of the unbearable eternal present and expression of love and courage in spite of overwhelming suffering, Grace evokes the heroic potential of her generation.

Sarah Kane's *oeuvre*, though it consists of just five stage plays and one screenplay, remains unparalleled among her cohorts for the singular way in which it generated the energy of the post-Thatcher period. These works stand alone in a radical stylization that seems to call for new forms of theatricality and performance, and in their use of classical models, such as *King Lear* or *Oedipus Rex*, as reference points for a neo-mythical generational vision. Kane's plays deal with a politics of the soul. Their consistent thread centres on a young woman who transcends boundaries, overcomes victimization, offers love, and through this love effects change in a violent and hate-filled world. Kane's plays connect with other in-yer-face plays in the evocation of a troubled generational psyche and the certainty of emptiness at the centre of subjectivity. Regrettably, Kane's early death prevented her from fully exploring or developing the provocative paradigms set out in her first three plays. Nonetheless, they remain significant for an idealism based in a particular historical moment and inextricably linked to the ferociously cruel conceptual worlds in which it is formed.

Mark Ravenhill's critique of commercialized culture

While Kane focused on pain, Mark Ravenhill, whose first play, *Shopping and Fucking*, was produced by Out of Joint and presented at the Royal Court in 1996, focuses on pleasure. Ravenhill's plays, however, reflect a peculiarly joyless and even lifeless kind of pleasure associated with, as the title of his first play makes clear, the escapism of habitual consumerism and casual sex. Ravenhill, in contrast to Kane, could be said to articulate a politics of soullessness. His plays not only expose audiences to explicit scenes of sex, but also show those acts as empty of emotional connection. Emotional emptiness points towards a deeper void in isolated lives that draw nothing from social, moral, or historical sources of meaning. The characters through which Ravenhill represents his generation take affluence, freedom, and even privilege for granted. They move in a fast-paced and brightly lit urban landscape evoked in Max Stafford Clark's production through neon signs, video, and rhythm-driven techno music. At the same time, they show a sense of loss in the lack of identity, relatedness, and values not rooted in business or the profit motive. As Rebellato states in his introduction to Ravenhill's *Plays I*, these characters test 'whether there is anything left in our lives together that cannot be bought and sold' (xi).

Shopping and Fucking employs social realism's focus on observable behaviour in a series of sequential but disconnected episodes to explore the lives of a group of young people. It begins with Mark, Lulu, and Robbie, a family-like group, having a meal of take-out food. The three recall their meeting with the 'shopping story', a quasi-romance narrative about Mark striking a deal with a 'fat guy' in a supermarket to 'buy' Lulu and Robbie (5). The play's action hinges on Mark's need for separation from Lulu and Robbie, to free himself from the anaesthetizing effects of drug dependency and a set of relationships expressed in terms of objectification and commerce.

Mark's subsequent quest leaves Robbie and Lulu to survive on their own. The play splits into two tracks, as Mark and the couple form new associations. Seeking sex without attachment, Mark nevertheless becomes involved with Gary, a 14-year-old rent boy. When the teenager describes being abused by his stepfather, Mark's impulse for caring arises in response to Gary's powerlessness and pain. Mark attempts to set a new pattern with this relationship, but Gary relates to Mark only in the context of a single-minded desire for a father. Unable to express his need in language free of internalized oppression, Gary asks to be 'owned', and Mark therefore takes him to rejoin the quasi-family group back in the flat.

Lulu and Robbie meanwhile have been initiated into the world of economic necessity. Robbie has been hired and fired from a fast-food service job, and Lulu has been stealing frozen dinners. Through an audition for what she thinks is an acting job, Lulu has become involved with Brian, a gangster obsessed with the father–son relationship in the *Lion King* movie. Moved by her Chekhov monologue about work and existential anguish, Brian hires Lulu to sell a batch of Ecstasy tablets. Lulu fails this trial assignment by giving the drugs to Robbie, who gets high on some and gives away the rest. Brian threatens them with torture, and mobilized by the threats, Robbie and Lulu have begun to make money performing phone sex.

The conclusion of the plot involving Mark and Gary ends in a mystery hinting at tragedy, while the conclusion of the one involving Brian, Lulu, and Robbie ends in comedy. When Mark returns to the flat with Gary, the resulting discomfort resolves itself in a fantasy game that culminates with both Robbie and Mark inflicting violent anal sex on the blindfolded Gary. Gary pleads for a father and repeats his fantasy that his father will sodomize him with a knife. The end-game implications of this fantasy frighten Robbie and Lulu enough that they break off the game, but Mark, acting alone, leads the blindfolded Gary out. Then Brian arrives, and Robbie and Lulu hand over their phone sex earnings in payment for the lost drugs. Brian's unstable persona, however, mutates from gangster to evangelist, as he preaches a gospel of 'get the money first' and 'money is civilization' (87). Satisfied that they have learned this lesson, he returns the money to them and departs triumphantly. The final scene re-establishes stasis as Mark, Lulu, and Robbie share food from individually packaged meals while Mark tells a new version of the shopping story.

The parallels to Osborne's *Look Back in Anger* suggest a conscious attempt at a new generational manifesto. Like Jimmy Porter, Alison, and Cliff in the 1957 play, Mark, Robbie, and Lulu play out their quest in the form of games and stories. The cultural context for their relationships, however, differs markedly. The games they play use the vocabulary of commerce rather than the animal metaphors of Jimmy and Alison, suggesting an economic rather than a natural frame of reference. While Jimmy's rebellion targets his wife's upper-class, patriarchal father, the characters of *Shopping and Fucking* acknowledge no authority and thus have no starting point for rebellion. Instead, they all seem to be looking for a father or an authority figure. In the absence of authority, they live without rules or expectations. Sexual acts take place openly, and gender codes fail to address a *ménage à trois* in which neither male desires the female. Drug use, stealing, and sex with a young adolescent occur without apparent concern for consequences. Despite Lulu's assertion that her 'instinct is for work' (12), she and Robbie do not work until coerced by Brian's threats into earning money. The only threat comes from displeasing Brian, but his ultimate transformation from tough gangster to avuncular preacher contributes to the disorientation and lazily anarchic quality of their environment rather than countering it with the cause-and-effect certainty he first seemed to represent. In this disorienting world, all three of the younger characters express the desire to be 'owned', depending on a more powerful person to make decisions. The image of the willingly blindfolded Gary is emblematic of a generation coming of age while refusing to see, let alone interact with the world outside their personal obsessions.

The play presents to audiences the world of explicit and casual sex that Gary cannot face. In deliberate contrast to the vision of gay consciousness plays like Jonathan Harvey's *Beautiful Thing* (1993), sex is utterly without redemptive value. The freedom to exhibit sexual orientation and activity has not liberated individuals or given them a path to establishing meaning. In the world of *Shopping and Fucking* sex challenges intimacy rather than expressing it. Like money, sex is a function of power that turns people into objects of exchange and teaches them complicity in their own oppression. Lacking a connection with love or care, and lacking political power as a transgressive act, sex lacks creative potential for the characters. Rather than serving as an area of discovery, practice of love, or means of psychological regeneration, it has become merely a habit that fosters disgust and self-annihilation. Only Gary insists on meaningful sex, but his masochistic quest arises from a deep need to enact the trauma he has suffered.

Mark's break with Robbie and Lulu signals his need to find a touchstone of reality and potential meaning within himself. In spite of the fluid and boundary-less culture of which he is a part, Mark attempts to establish a traditionally masculine unitary identity. Seeking boundaries, he first tries to shed his dependencies in a drug rehabilitation unit. His sexual impulses, however, violate the rules and result in his expulsion back into the world of choice. As Robbie indicates in an anecdote about a fast-food customer who goes berserk because he cannot decide whether or not to have cheese on his burger, choice is maddening without a clear sense of self from which to exercise it. Once on his own, Mark finds his quest for selfhood lost in oppositions between personal and impersonal or powerful and weak that perversely transform into each other and confound his efforts at identifying options. Mark's quest, though enacted in different terms, comes to much the same conclusion as that of Jimmy Porter. He finds unitary identity impossible except, perhaps, in the act of causing death, and uneasily resumes old habits at the end.

The play's themes of identity and connection turn on the impossibility of knowing others in ways that go beyond simple observation of observable reactions. As Gary says to Mark in one scene, upon observing his erection, 'What's going on in your head? I mean, I can see what's going on in your pants, but what's in there?' (3) The characters attempt to communicate through narratives of what they have experienced when apart from the others, but none can corroborate the others' stories or validate their experiences. Truth cannot be separated from fiction. Mark's story of sex with a 'famous person' (69) brings a flash of derisive anger from Robbie, who sees it as evidence of Mark's deceptiveness. Gary accepts the story as truth but sees in it only that Mark has had sex with a woman. Even the stories that seem to be true fail to elicit understanding in those who hear them. Lulu vividly describes to Robbie a random murder she witnessed in a convenience shop, which left her spattered with blood and haunted with guilt, but he does not understand how this event has changed her sense of self. The lack of shared contexts impairs understanding of others' stories, as can be seen when Brian relates the plot of *The Lion King* as though it were a completely original story, while Lulu anticipates its incidents by drawing from a cultural reservoir of which Brian seems completely ignorant. The narratives, according to Robbie, are 'little stories' – personal crutches used to 'get by' (66). Echoing Jimmy Porter's famous complaint that 'there aren't any good, brave causes left' (Osborne, 1965:104), Robbie states,

> A long time ago there were big stories. Stories so big you could live your whole life in them. The Powerful Hands of the Gods and Fate. The Journey to Enlightenment. The March of Socialism. But they all died or the world grew up or grew senile or forgot them, so now we're all making up our own stories. Little stories. (66)

Stories fail to provide meaningful connection because storytelling, like sex, has become an element of commerce. Robbie and Lulu's telephone sex business is storytelling for hire. Brian has found inspiration in the narrative of a commercial animated film. Mark, Robbie, and Lulu identify their relationship and their shared generational position through a narrative of commercial exchange told in three versions. The first is Mark's story of buying Robbie and Lulu in a supermarket – a fantasy that epitomizes their relationship. The variations that follow indicate the ways in which the relationship is changing. Robbie signals conditional acceptance of Gary when he agrees to tell the story as part of a sex-for-hire game with Gary. He specifies Gary as the one who is bought, but an agonized Gary substitutes his own narrative of finding his father and receiving definition through his act of violent finality. Gary's narrative articulates self-definition and desire, but communicates loss and futility.

Gary's despair convinces Mark to make a choice. In the final moments of the play Mark tells Lulu and Robbie the final version of the group's narrative. Giving the story a sci-fi twist, he tells of buying a slave creature and then setting him free. The decisive moment has occurred offstage, but the profoundly disturbing image of Mark's dispassionate resumption of his place soon after leading away the blindfolded Gary prompts unanswerable questions. Gary's story has taken on central importance, but the play does not reveal its conclusion. The gap in dramatic action exposes an awkward desire for traditional resolution but an inability to represent such closure. The heroic role Mark gives himself in the fiction of freeing the slave not only indicates that he cares for the desperate teenager but also highlights the conflict between caring and freedom. The play's final moment, when the three flatmates share individually packaged meals in silence, shows a reversion to habit, in which the quickly consumed and disposable products of material culture allow them to continue a communal life without decisive commitment.

Shopping and Fucking deals with the legacies of the Thatcher period, when liberalization of laws and attitudes about homosexuality and the targeting of gay men as a market sector coincided with Thatcher's restructuring of the economy, education, and social services to reflect her market-based philosophy. The play contextualizes the personal and political disengagement of Ravenhill's generation in the conditions created by a culture of consumption that has replaced parental and other authority in structuring individual identity and choice. In the absence of guidance and support, they have failed to establish personal or collective identity. Without legal or social constraints that at least provide a basis for rebellion, they find it difficult to make conscious choices or find meaning in those choices. They long for family and attempt to mimic its security in their quasi-family group, but cannot create this basic form of stability, comfort, and protection. They seek fathers, but find only fellow orphans and the buffoonish marketer of ersatz gospels. Subject to a constant barrage of random events, they are in danger of being overwhelmed by chance encounters that they cannot incorporate into any larger framework, such as Lulu's experience of being splashed with blood in a supermarket. They are unable to visualize or construct a communal history, and this absence negates the possibility of meaning in acts such as Robbie's free distribution of the Ecstasy tablets as surely as it negates the possibility of taking a meaningful action to save the troubled rent boy.

The characters' disability has roots in the previous generation's neglect and abuse, vacuous Hollywood myths, the ready availability of chemical agents of escape, and a pervasive market economy that is neither forgiving nor rational. Their only shared experience and verifiable reality exists in the present and takes as its substance unwholesome packaged food items. Ravenhill thus reveals the postmodern dilemma of ostensible pleasure without guilt that negates conditions of choice, the absence of a stable narrative of identity or history, and the repetition of present moments through habit. The play suggests that members of the post-Thatcher generation are unable to act upon the ethical and political issues of their time because of their dependence on material things and tendency to objectify relationships, their self-indulgence that has annihilated self-respect, and their fear of 'big stories'. They cannot articulate a shared reality or confess their unique failures and acts of destruction. A world lacking moral boundaries deprives them of risk, tragedy, and joy.

In his subsequent *Faust Is Dead* (1997) and *Handbag* (1998), Ravenhill explored issues of postmodern identity introduced in his first play, and then went on to write *Some Explicit Polaroids*, produced in 1999 by Out of Joint and first performed at the Theatre Royal, Bury St. Edmonds. *Some Explicit Polaroids* stands out as the most explicitly political of Ravenhill's early plays. In it Ravenhill confronts the apologia for disengagement he presented in *Shopping and Fucking* and at the same time rejects 1970s-style leftist politics as a framework for social change. Contemplating the forms both personal relationships and political engagement might take, *Some Explicit Polaroids* draws a clear dichotomy between 1970s politics and post-Thatcher politics. Through this contrast Ravenhill addresses the challenge of contemporary conditions for personal commitment and for political thought and action.

Written in realistic style, the play conveys its ideas through straightforward dialogue and debate between pairs of characters against a backdrop of nightclub music and technical gadgetry situating the action within trendy urban culture. Nick represents 1970s activism. Just released from prison, where he was sent in 1984 for stabbing a man he considered a corporate criminal, he still wants to believe in his youthful commitment to 'destroying the rich' in order to 'wipe out a thousand years of suffering' (263). New developments, like phone cards and children selling drugs, confuse Nick, but changes in the *zeitgeist* create even greater disorientation. He visits Helen, the woman with whom he shared youthful radicalism and a romantic relationship, but is disappointed by her accommodation to post-Thatcher conditions. She serves as a local councillor, working on projects like improving bus service so that people on the council estates can 'get to the shops' (237), and she has parliamentary ambitions. Helen greets Nick coldly, refuses to take him in, and bitterly turns back his criticism of her present choices by challenging him to name one thing their radical activism accomplished. Nick then gets involved with Nadia, an erotic dancer with an abusive boyfriend, whom he encounters by chance as she is being attacked by her boyfriend. Nadia offers the shallow idealism of pop psychology, chattering continually about the importance of emotions but refusing to acknowledge her blood and bruises or to protect herself against the boyfriend, because she cannot face the reality of pain, anger, and fear.

Nadia and her two friends, Tim and Victor, provide a composite picture of the younger generation of post-Thatcher Britain. They live in a world of unfettered self-indulgence but seem to have misplaced their selves. They, even more than the characters in *Shopping and Fucking*, seem trapped in consumer culture that negates choice. Nadia, Tim, and Victor deny their pervasive sense of emptiness through the construct of 'happy world' (269) – a reference to their cohort as Generation Ecstasy. They reinforce this ironic illusion with perverse pride in their consumption of junk food, pornography, and other products of 'trash' culture. Tim, whose immune system has been compromised by HIV, exists in the limbo of life-prolonging drugs. Tim's lover Victor, an economic migrant from Eastern Europe, hopes to market his 'fucking fantastic' body and become 'a huge porno star' (239). Tim plays a mock paternal role with Victor and Nadia, offering them sweets and kisses while coaxing them to disconnect from anger and escape into the happiness illusion. Nadia hides her bruises under make-up as she claims, 'we're at peace with ourselves...we're our own people' (273). Victor condemns Eastern Europe and its socialist politics as ugly and eagerly develops his own routine of erotic dance performed in a cage. The three are not as insensible as they first appear; in odd moments they permit glimpses of the potential for love and insight under their veneer of denial and casual objectification of self and others. The turning point for this trio occurs when Tim decides to discontinue his life-prolonging drugs. His illness and death force honesty to the surface in their relationships.

Although it initially seems to offer a pessimistic view of political action, *Some Explicit Polaroids* ultimately reveals significant potential for change, at least on a personal level. Each character experiences a decisive moment that alters his or her views or life direction. Tim initiates the chain of personal changes by refusing his medication, in order to 'know where I am' (288). Choosing objective reality over happy world, he faces the cause-and-effect certainty of dying from immune deficiency. Through his death, Nadia and Victor acknowledge love and loss, and discover an ideal of Tim that persists beyond his death. Strengthened by this ideal, they risk experiencing life without the 'happy world' myth.

Nick's decisive moment occurs when he finally confronts Jonathan, the man he injured in his attack. The enormously wealthy Jonathan has become an international philanthropist who funds schools and hospitals in Eastern Europe. He is, as Nadia observes, 'at ease' with his authority (290), and he speaks of his conscious choice to embrace the excitement of 'being swept along by the great tides and winds of the market' (293). Despite his position and power, he still has thoughts of retaliation against Nick, but when he sees Nick as he really is, beaten down to the point where he often wishes to be back in prison, he offers to help him. Both men acknowledge nostalgia for the defining conflict of the past. Nick tells Jonathan, 'It was much easier. Before. When I hated you. I knew where I stood' (311). Jonathan admits, 'I think we both miss the struggle' (*ibid.*). Jonathan projects an interesting ambiguity: although his self-possession and involvement in humanitarian work create him as a sympathetic figure, he is the only character who remains alone throughout the play. He shows little capacity for personal connection, but does serve as a helpful model for Nadia, who finds hope in the way he has overcome the victimization forced on him by Nick's attack.

Once Nick has made his peace with Jonathan, he and Helen resume their relationship, and Helen vows to try to reinvigorate his spirit. Thus, in *Some Explicit Polaroids* the older characters overcome false concepts of one another, while Nadia and Victor discover hope in the final image of their friend Tim. The play concludes with a fully formed narrative that creates a tentative paradigm for political action. The older characters create this paradigm, and by the end of the play their lives have become separate from those of the younger characters. It is not clear whether the younger characters will be able to construct the basis for political action from the material of Tim's death, Jonathan's commitment to the market, and their own exposed emotions. However, these factors have combined to push them out of the nest of habit, and Tim's death has given them a common reality and a memory that they will take with them into the future.

The new politics articulated in *Some Explicit Polaroids* eschews anger and the clash of wills. This politics grounds political agency in non-heroic ideals of action and a sense of reality-based limits. In accepting limits, it validates compromise and values small gains. Ravenhill's political vision, like Sarah Kane's, proceeds from the experience of love rather than opposition or enmity. Both playwrights show love as difficult to find or choose, given an uncertain sense of reality, lack of sustaining role models or narratives, and the ubiquitous temptation to substitute easily obtained consumer goods for the demands of identity, choice, and relationships. Both Kane and Ravenhill include in their plays members of their generation who are denied love by sexual exploitation and driven towards death by the impossibility of love. At the same time, their plays suggest that love is the only force strong enough to bring about change on a personal level, and imply that participation in the world beyond the self begins with personal change.

Kane and Ravenhill stand out as the most influential playwrights of their generation, and their plays provide an intense focus on broad philosophical themes that attempt to define their generation politically. However, they do not define the entire cohort of post-Thatcher playwrights or fully demonstrate the range of in-yer-face plays. As Aleks Sierz has noted, the Royal Court's 1994–1995 season, which was a conscious effort on the part of the Court's artistic director Stephen Daldry to attract younger audiences through new writing, introduced not only Sarah Kane but also Joe Penhall, Judy Upton, and Nick Grosso. The following several years saw the emergence of additional young playwrights, including Jez Butterworth, David Eldridge, Rebecca Prichard, Abi Morgan, Martin McDonagh, and Stella Feehily at the Royal Court, the Bush, the Hampstead, the Soho, and the National theatres.

Several of these young playwrights, including Penhall, Upton, Eldridge, Prichard, and Feehily, used variants of social realism to address the conditions of post-Thatcher Britain with topical images of a generation under threat. They focus on specific issues, but the characters and situations of their plays have a resonance that goes beyond the particularities of their stories and argues against Thatcher's dismissal of society. They show a concern over issues of rationality, character development, and maturity, within the context of the failures of families and society to provide effective structures of protection, guidance, social integration, and opportunity. Use of actual locations, linear plots, and vividly created, highly individualized characters work together to evoke the aspects of society dismissed by Thatcher. They appeal to public conscience to recognize and help the victims of ruthless cuts in social welfare programmes.

Marginality and conscience

Joe Penhall's first full-length play *Some Voices*, presented at the Royal Court just a few months before Sarah Kane's landmark *Blasted*, centres on a pair of brothers in West London. The two demonstrate contrasting sectors of their generation – those who lack the ability to be responsible and those who carry extraordinary burdens. The older brother Pete, who is in his thirties, leads a strenuously disciplined life as the owner and operator of a café. Motivated by the need to live down his father's failure to keep the café in business, Pete works very hard to succeed, exemplifying the Thatcher ideal of self-reliance. The younger brother, Ray, a schizophrenic in his twenties, is discharged into the care of the community when he has spent the maximum allowable time in the hospital. As Ray's only relative, Pete becomes, in effect, the community. Ray soon stops taking his medication and disrupts his brother's life with his bizarre behaviour. Rather than finding support in the wider community, Ray connects with others who need help. He has an affair with a pregnant woman and violent encounters with her abusive boyfriend. In episodes that climax with a threat to burn down the café, Ray nearly destroys Pete's equilibrium. Strangely, he manages to liberate the abused woman, who leaves the abuser and Ray as well, going off to live alone. Pete, despite the stress of dealing with Ray, remains involved with his brother and attempts to find him a permanent living situation.

National life, as exemplified by London in this straightforward play, includes people with extreme needs. Thatcher's insistence that such needs should be taken care of within families has resulted in a dysfunctional brother threatening to overwhelm the functional one, endangering his stability and productivity. With a subtle duality of perspective, *Some Voices* allows the audience to see schizophrenia both realistically as a mental illness that affects individuals and metaphorically as an indicator of social breakdown. In a note accompanying the play in an early anthology, Penhall reveals that the character of Ray is based on a friend whose 'confusion, hypersensitivity, depression and wild, antic humour had always seemed an appropriate enough response to life' (Edwardes, 210). Similarly, its focus on an individual who needs social and medical services highlights the gaps in those services caused by inadequate funding and bureaucratic rigidity, while the broader implications of an inadequate system are reflected in the struggling residents of a stressed urban community.

Penhall explores a similar problem while developing an argument about the generational divide in *Blue/Orange*, first performed at the National Theatre in 2000. This play pits a young psychiatrist-in-training against a seasoned one as the two disagree over the best course of treatment for Christopher, a young Afro-Caribbean man diagnosed with borderline personality disorder. Although Christopher, who was hospitalized after making a public disturbance, has reached the end of his allowable time, he continues to make delusional statements, claiming the oranges in a fruit bowl are blue, that Idi Amin is his father, and that vaguely defined others are harassing him. Bruce, the younger psychiatrist, feels that Christopher needs longer hospital treatment and wants to revise his diagnosis to schizophrenia to permit that. Robert, the supervising psychiatrist, wants to follow rules and release Christopher to the community. Bruce argues that Christopher has no known friends or family and that the newer antipsychotic drugs with few side effects, which he would perhaps tolerate as an outpatient, are too expensive to prescribe for a patient without private resources.

Although the two psychiatrists begin the play by addressing one another cordially, the power differential between them becomes increasingly obvious. Bruce bases his case for continued hospitalization on empathy with the patient and observation of his behaviour. Robert uses two languages – the language of power and the language of abstract idealism. He informs Bruce that he will take over the case, and when Bruce threatens to go to the authority overseeing the hospital, replies, 'I *run* the Authority' (48). At the same time, he couches his use of power in idealistic terms. He interprets Christopher's delusions as analogies describing the oppression of the black minority in England. He insists that the hospital is 'an anachronistic Victorian institution that should have been closed down years ago' (90). These justifications mask a deep cynicism, evident in his deliberate manipulation of the literal-minded Christopher, which leads him to lodge a complaint that takes Bruce's words out of context and brands him a racist.

While it establishes a clear opposition between the sympathetic and logical Bruce and the hypocritical and egotistical Robert, who punctuates his pronouncements with a braying laugh, the play does not clearly establish the rightness of either medical opinion. Robert's assertions regarding social conditions that produce disorders of personality and thinking, and his reminder that a longer hospital stay could actually make the patient worse, carry enough validity to cast some doubt on Bruce's diagnosis and recommended treatment. In fact, as Bruce continues to insist on this single line of action in the face of a variety of arguments, he begins to seem unbalanced and even obsessed with controlling Christopher. When his frustration over the case reaches a climax, he vents it in a violent and ill-considered verbal outburst that borders on irrationality. Christopher wavers from one side to the other. He initially trusts Bruce, but switches loyalty when Robert sees him privately, and ends by lodging the complaint against Bruce. At times Christopher seems like an innocent being used by those in authority; at times he seems like a canny operator allying himself with the more powerful agent in this situation in order to gain control.

The outcome of the play, though predictable, does not establish a sense of reality or truth. It leaves audiences with conflicting feelings about the failure of Bruce and Robert's professional relationship, the state of Christopher's mind, and the place of idealism in addressing social problems. Bruce ends up not only defeated in his attempt to continue hospital treatment for Christopher, but also relieved of his responsibilities and facing the ruin of his career. Bruce's mentor has betrayed him and annihilated his hopes while exhibiting breathtaking vanity. Bruce, though he puts his patient's welfare above his own, shows an over-confidence in his judgment that signals his customary power. Christopher, in common with his imagined father Idi Amin, may be irrational in any cultural context or may be the victim of distorted perceptions that prevent him from being understood in rational terms. Finally, Robert's use of idealism and Bruce's corresponding reliance on psychiatric categories call into question the value of either approach in healing or problem solving. The psychiatrists' interaction most clearly points to the tendency of institutions to serve the interests of those in charge of them rather than the needs of those served by them. Bruce's final vow to 'lodge a Complaint with the Authority' (115) may be seen either as an exercise in futility or a note of defiance and hope.

Blue/Orange, like *Some Voices*, highlights a lack of commitment to caring for those who need help and protection, while making a case for the necessity of community. It goes further, however, in questioning the competence and disinterestedness of the institutions and structures of authority charged with responsibility for the mentally ill. The psychiatrists, who quickly abandon rational argument and engage in a raw power struggle over Christopher's case, demonstrate the inadequacy of a purely institutional response. Since Christopher lacks even a rudimentary family, Thatcher's solution cannot be applied in his case unless the psychiatrists were to take him at his word and pack him off to Uganda to live with the man he claims to be his father. As it is, he will be released to a community that has made no provision for supporting him. Christopher, like the disturbed Ray in *Some Voices*, will inevitably and ironically repeat his disruption of the productive segment of society, as he did when he was taken in hospital after making a disturbance in a street market. Both of these Penhall plays thus argue that the mercantile structures of central concern to Thatcher actually depend on a community system of care and protection for their optimal functioning.

Serving It Up by David Eldridge, which premiered at the Bush Theatre in 1996, focuses on the harmful potential of a single individual and, by extension, the harm inherent in a refusal to accept reality. Eldridge places this individual within the realistically presented socio-political context of a white working-class community in London's East End. Sonny, a young unemployed man in his late teens, bases his identity on his family and East-End culture, but does not accept the reality of either. Lacking curiosity, ambition, and introspection, Sonny earns some cash by dealing drugs. He anaesthetizes himself with drink and drugs, engages in the random violence of street fighting, and expects an early death. Only racial hatred rouses him to passion. Though he cannot explain why he hates 'darkies and Pakis … Kurds … and fucking Greeks' (52), he expresses alarm at their presence in England. He refuses a political identity, grumbling that 'politics is crap' (51), but shaves his head and lays claim to patriotism as a dubious cloak for loutish drunken behaviour when England fails to win the World Cup. Sonny's boyishness draws sympathy to him in spite of his verbal and physical aggression.

Sonny and his friends feel their lives diverging. Their common identity breaks up as his friends make choices that take them in new directions and Sonny finds himself increasingly alone. Ryan, a friend who went away to university, has relocated permanently and is preparing to marry. Ryan and Sonny used to share an avid interest in cricket, but when Ryan's mention of his fiancée's name reveals that she is non-white, Sonny ends the friendship in a surge of insults. Wendy and Teresa, two childhood friends, now socialize with Sonny only because of interest in his friend Nick. Sonny regards Wendy as his girlfriend, but when he tries inarticulately to communicate this to her, she responds that she will not tolerate such abuse. Nick, Sonny's only close friend, spends much of his time at Sonny's house to escape a violent father. Nick reads books, watches educational television, and gets a job with the local council. These choices strain their relationship, but the decisive break comes when Nick inadvertently reveals that he has been having an affair with Sonny's mother. This revelation demolishes the constants on which he has based his identity, and Sonny sinks into a profound, silent rage. His climactic act of violence occurs offstage, but cuts him off so completely that he is unseen at the end. The final scene shows Sonny's parents and Wendy and Teresa sitting in separate pairs, stunned by what has happened. Wendy, who had taunted Sonny with her preference for Nick, feels that the tragedy is her fault. When Teresa calls Sonny a monster, Wendy responds, 'We're all monsters' (84). Sonny's mother Val cannot articulate her feelings about what has happened; while she cries silently, she obsessively eats the cake that she has tried unsuccessfully throughout the play to feed to the men in her life.

Serving It Up highlights a skinhead's hidden insecurity, which holds greater potential for harm than his visible belligerence. Like many other in-yer-face plays, it calls attention to the failure of boundaries, draws a contrast between productive and non-productive members of the post-Thatcher generation, and focuses on the impact a disordered person may have on a community. Sonny founds his identity, racism, and belligerence on the ostensible rock of his family, but this rock has been turning into sand. Val cannot get her husband to eat any of the cake she is constantly serving up, and so turns to Sonny's friends, of whom Nick is only the latest. The line Sonny believes to define his family's unity has been crossed many times, but Sonny has refused to face the reality. His overwhelming need to preserve his illusion makes him turn his best friend into an enemy. The ensuing violence robs the community of a promising and productive young man, brings the presumed incarceration of another, and places a burden of paralyzing guilt on those closest to them.

The young characters emphasize the variety of personal and political attitudes among a generation. The range of experiences and orientations illustrates the challenge of definition even among people who share a similar background. It shows how divergence can create conflict as people who were once friends make different choices and ally themselves with opposing interests. The focus on differences links the personal and the political, but none of the characters shows energy or interest in changing society. More than one of the young characters comments on children playing in the park and speaks nostalgically of childhood. They would like somehow to recapture for themselves the unblemished hopefulness they see in the children. Though only in their teens, they are already weary of choice and responsibility.

Girls and power

Margaret Thatcher, as Caryl Churchill made clear in *Top Girls*, exemplified an individualist philosophy antithetical to feminism that encouraged women to compete for money and power and suggested that toughness and insensitivity were the qualities that would enable them to compete successfully. *Ashes and Sand* by Judy Upton, which premiered at the Royal Court Theatre in December of 1994, focuses on tough young women who grapple for money, recognition, and power in a deprived and decaying city. With its disturbing picture of violence perpetrated by young adolescents, *Ashes and Sand* bases itself on the actual phenomenon of girl gangs. It explores what Aston, in *Feminist Views on the English Stage*, has identified as the Thatcherite discourse of 'girl power' associated with the pop music group Spice Girls, in relation to 'communities of disadvantaged young women in the 1990s' (6).

A deteriorating boardwalk in the seaside resort of Brighton provides the territory roamed by a gang of teenaged girls who have little guidance, few prospects, and a great deal of unfocused resentment. Uninterested in school and pessimistic about future opportunities, they spend their time hanging out in arcades and shops. Hayley, the gang leader, lures men so that her friends Anna, Jo, and Lauren can move in and steal their wallets. Hayley hoards the stolen money and nurtures the group's fantasy of an escape to Bali. Two police officers, Daniel and Glyn, keep an eye on the boardwalk, but Daniel's shoe and make-up fetishism opposes him to the norms he is hired to enforce, and his own desire for escape draws him into a dangerous closeness with the girls.

The gang members flirt with Daniel and engage in rivalry over him, but when he is arrested for shoplifting shoes, the gang splits. Hayley allows her obsession with Daniel to dictate her actions. She persuades Anna, who works at the shoe shop, to give a statement exonerating Daniel, and then tries to communicate her love for him through a locked door he refuses to open. Jo and Lauren steal Hayley's hoard of money, squander it on drugs and clothes, and then find themselves in serious trouble when they kill a man in the course of a robbery. They hope Daniel will save them by providing an alibi, but he has managed to obtain a transfer to Gibraltar and is planning to leave. The girls reunite when they think Daniel has betrayed them. They break into his room and attack him when he returns, turning him into an object of sacrifice in a ritualistic enactment of violence.

The desperate energy of teenagers who sense that they will soon become casualties of the perilous transition from childhood to adulthood fuels the play's frenetic pace. Brief scenes show bursts of activity that often culminate in violence. The empty environment suggests deprivation, especially in its absence of the kind of social boundaries that should inscribe belonging and set limits to the behaviour of the girls. In the absence of concerned adults, the girls find in their group a sense of belonging, but competition among peers disrupts the group's unity. The girls temporarily seem to triumph over some constraints of gender and class, but their lawless aggression does not enable them to escape the environment or transform the conditions of their powerlessness. The burden of control falls on the police who are also trapped in the deprived landscape. Daniel's own orientation towards crossing boundaries makes him unable to maintain limits for the girls, and the sexual magnetism he half-consciously exerts on them further complicates his role. With his sexuality absorbed by his fetish, Daniel desires no more than casual flirting, but Hayley's fierce sexual energy compels her to take some action – if not to possess, then to destroy.

The pervasive boundary crossing blurs the separation between the police and the law-breaking gang. The original production's fluid staging and use of mirrors throughout the set merged specific spaces into one another as the relationship between Daniel and the girls became more confused. The absence of a defining framework allows the girls and Daniel to become entangled in emotional trajectories that lead to disaster. The second policeman, Glyn, with an enabler's blindness to the obvious, fails to put a stop to Daniel's escalating involvement with the girls. The furious culmination of the play's action shows the gang's increasing capacity for irrational violence. The youthful energy and dreams that have evoked sympathy and hope for the girls seem, by the end, cruelly futile.

Upton's focus on girls points to gender relations in the post-Thatcher generation. A kind of gender consciousness has impelled the girls to band together to enhance their power and turn the tables on male predators. They accomplish this, however, only by using their apparent powerlessness as a trap. The vaguely formed desire for escape highlights confinement not only in their deprived environment but also in structures of gender and class. The girls' fantasy of going to Bali, however, implies an exotic vacation rather than a break with the assumptions and habits that constrain the lives they can imagine for themselves. The gang's initial solidarity splinters when confronted with the pressure of individual desires. Hayley's leadership fails because she lacks the consciousness and discipline to sustain commitment to a common goal. The lack of commonly understood ideals or even ideas deprives the girls of the means to generate new leadership or articulate new goals. As a result, the girls who spend the hoard of money that represents their collective power squander it impulsively. The girls' final orgy of violence against Daniel, whom they target as an enemy without understanding how he has allied himself with them, also does nothing to alter their powerlessness.

Ashes and Sand confronts audiences with the perversion of adolescent energy and the waste of young lives amid casual and irrational violence. It presents a naturalistic view of a group of young people whose destructive behaviour arises from norms of amoral competition coupled with the lack of adult guidance, the deprived conditions of the community, and the inevitable frustration of unrealistic desires. In an afterword to the play, Upton described being angry when she wrote *Ashes and Sand*; however, the play communicates resignation rather than anger by focusing only on the effects of what it seems to suggest is a cause–effect sequence. The strength and savage pleasure the girls reveal in their communal violence suggest that their collective power will never be employed towards creating better lives. The lack of resolve on the part of the authorities points to a broad willingness on the part of society to tolerate the waste of young lives that are not apparently valued by families or home communities. Upton thus shocks the audience by showing this extreme social dysfunction without indicating any solution.

London's East-End borough of Hackney, balanced uneasily between the gentrification efforts of wealthy newcomers and the efforts by its poor population to simply survive, provides the setting for Rebecca Prichard's *Yard Gal*. This play again confronts audiences with a girl gang. Commissioned by the Clean Break Company and first performed at the Royal Court Theatre in 1998, the play's two characters, Marie and Boo, give a riveting account of their involvement with a 'posse' of girls who deal drugs, steal, and fight. These girls have no real home or parents, and abandonment has created a kind of freedom as well as constant struggle. Boo ran away from a children's home, where the caretakers 'just see it as a job' (14), while Marie fled a violent father. The posse, a female equivalent of the Jamaican–British gangs known as yardies, gives members an identity, a sense of belonging, and a form of power. Though Boo is black and Marie white, they both identify with Jamaican–British urban street culture and use its distinctive language. They tell of getting a tattoo as an inclusion ritual, styling one another's hair and exchanging clothes, and fighting with another gang. They own almost nothing, and their only place of retreat is a vacant flat with broken windows, a trashed interior, and a smell 'like somebody died' (34). Most of the time they are on the streets.

The security of numbers provides some protection, but violence always shadows the yard girls. One falls to her death after drinking a bottle of rum and then attempting to jump onto a neighbouring balcony. Violent clashes with a rival posse bring injuries and arrests. Loyalty to friends is the girls' primary value, but loyalty precipitates the end of the friendship at the centre of the play. Boo demonstrates loyalty when she takes the blame for Marie's attack on a rival, but then she is sentenced to prison. Separated, the two gradually lose touch, though each continues to regard the other as her 'best friend . . . from time' (59).

Yard Gal makes no attempt to address issues of an entire generation, instead concentrating on girls at the farthest edge of society's margins. The portrayal of this often-invisible population has an authenticity related, no doubt, to its association with Clean Break, a company devoted to theatre work with ex-offenders. Despite its straightforward structure and simple trajectory of action, the play involves audiences in a complex emotional duality. On one hand, it evokes an exuberant joy in simply being alive that emphasizes the youth and lively personalities of the characters. At the same time, the play is pervaded by sadness. It reveals both the potential and the hopelessness of adolescents confronting the realities of a harsh urban life without familial or social support. The characters attract strong sympathy in spite of their behaviour. Given their lives of drug dealing and use, theft and violence, the outcome of the play is not surprising. What is surprising is the joyous side of these young women's relationships in spite of all the day-to-day obstacles they must surmount just to maintain a bleak status quo.

Yard Gal, like *Ashes and Sand*, offers a form of gender analysis. The girls in both plays seek to overcome powerlessness by banding together under a dominant leader, and in both the alliance fails when individual desires split the group. Ultimately, the gang threatens rather than supports the goal of survival. In *Yard Gal*, however, the girls tell their own story, and in the telling find a measure of power and meaning. The collaborative storytelling structure demonstrates the way narrative can be used to understand experience and construct meaning. Marie and Boo deliver interlocking narratives that make visible their friendship and demonstrate its centrality in their lives. Marie uses the friendship to support her tenuous entry into the realm of adult responsibility. When she gives birth, she names the baby for her absent best friend, and values the child enough to quit using drugs, look for a job, and stabilize her life. By contrast, Boo becomes lost and helpless in prison, finds herself unable to relate to life on the outside, but still clings to her memory of the friendship.

Telling their story, though important, does not enable either young woman to overcome her powerless position in society. At the end they ask permission to leave, indicating that the production of their narrative has been structured by others more powerful than themselves. This reminder of their limited autonomy counters the moments of joyful self-assertion in the narrative and leaves the audience to consider sombrely what the future for either young woman might be. Dramatized throughout in the form of direct address, the play constitutes a passionate plea to audiences to understand, think about, and value the young people of this blighted London area. Unlike other in-yer-face plays, it makes an implicit appeal to the collective social conscience. Boo and Marie argue for more effective support of abandoned youths and runaways by providing eloquent evidence that they are worth saving.

Duck by Stella Feehily, produced in 2003 as an Out of Joint and Royal Court co-production and first performed at the Theatre Royal, Bury St. Edmunds, was a latecomer to the in-yer-face moment. Set in Dublin, it pits the high-level energy of two young women against the stodgy confinement of their lower-middle-class environment. Both young women want to escape the unsatisfying lives they see their parents leading, and the desire for escape drives them to take risks. Nineteen-year-old Cat gets into trouble with her initial choices. She works as a waitress in a bar and lives with its drug-dealing owner, Mark, who treats her badly and taunts her with the nickname 'Duck' because of her large feet. She gets drunk and sets fire to Mark's jeep, and then begins an affair with a middle-aged writer she meets in the bar. Cat's best friend Sophie attends university and is doing well. Living at home with a nagging mother, Sophie gains a sense of freedom from a night out with Cat, but ends up with a broken nose after fending off an attempted rape by a pair of boys. Their different lives keep the two apart for some time.

Cat's limited options become clear when Mark learns that she destroyed his jeep. He nearly drowns her in an angry bathtub scene, and she leaves him. She returns to her parents' home, but they have no space for her physically or emotionally. She then goes to her middle-aged lover, but he proves unavailable for a permanent relationship and not interested in seeing her again after Mark breaks in and threatens him with a gun. Finally, she takes refuge with Sophie, and they decide to leave Dublin. Where they intend to go is not clear, but the two friends have cast their lot with one another.

Duck poses the question of whether the friendship with Sophie can save Cat from disaster or will merely doom Sophie. Cat takes an unlikely route to independence, first trusting her fate to the manipulative and violent Mark, then ending the relationship through the unconscious and dangerous act of arson. Cat's youth and vulnerability become literally visible in three nude scenes in which she first shares a bath with the aging writer, then with the abusive Mark, and finally alone at her parents' home with Sophie listening sympathetically to her account of events. Cat clearly needs help in her journey to maturity, but neither finds protection from men nor receives support from the older generation. Her only proven source of protection and support has been Sophie, and it is to Sophie that she turns decisively as an ally in escaping the environment that constrains both of them. Sophie's primary form of risk-taking is her association with Cat. In holding out the possibility, however small, that Sophie and Cat will create a viable form of girl power, this play eschews an appeal to conscience and even avoids suggesting that society has failed these young women. Instead, it implies that their problems have personal origins and may respond to a highly personal solution.

For the girls in these plays, banding together is a desperate measure rather than the freely chosen formation of a community. Their youth and inexperience make them incapable of participating competitively in a free market economy. They turn to one another because their families of origin and male lovers or friends have failed to provide them with support and protection, and their communities do not offer effective forms of care that might substitute for the absence or failure of families. Although their friendships and alliances offer respite from loneliness and despair, in practical terms they are the futile act of one drowning person clutching another. Upton, Prichard, and Feehily thus highlight the hopelessness of young women with little access to personal or collective power, emphasizing their potential and confronting audiences with the waste of this potential by a nation that does not recognize a responsibility to help them.

Marginality and power

Even more than the scenes of violence, what often shocked audiences about the in-yer-face plays was their cynical attitude towards wrongdoing. Acts that in reality would produce horrified condemnation were presented with casualness or even glee. In part, this attitude arose from the enjoyment of theatre by new playwrights' discovery of what Caryl Churchill once observed, that 'anything's possible in the theatre' (Klett, 1984). In addition, the amoral tone of many in-yer-face plays signals a rejection of notions of social conscience associated with the 1970s. The extremes of personal or social dysfunction presented in these plays challenge the optimistic assumptions and utopian aims of the social movements of the 1970s. Celebrating the human capacity for destruction intensifies defiance of communitarian social ideals and tests the limits of audience acceptance of extremes for their own sake, rather than as a means towards social goals.

In contrast to the concern for overcoming oppression and establishing justice typically seen in political drama, plays by Jez Butterworth and Martin McDonagh treat both victims and oppressors with caustic humour, deconstructing their opposition. In plays influenced by the hyped-up fantasy of action films, Butterworth and McDonagh treat society's margins as a dangerous but exciting playground in which desires that would be hidden in mainstream society surface and seek satisfaction. Location in the social margins removes the action from societal authority. In this ungoverned realm of savage competition, neither family nor friendship ensures loyalty. Butterworth's *Mojo* and McDonagh's *The Beauty Queen of Leenane* owe much to earlier models from popular culture – melodrama in the case of *Beauty Queen* and film *noir* in the case of *Mojo*. Both plays, however, throw these models off balance with moments of hilarity and the refusal to provide a sense of closure at the end of the play, showing influence from the blackly comic treatment of isolation and fear in much of Harold Pinter's work and in Philip Ridley's 1991 play *Pitchfork Disney*.

Butterworth's *Mojo*, which premiered at the Royal Court in 1995, uses the rock 'n' roll culture of the 1950s, a paradigm of youthful rebellion, to explore issues of youth and maturity. The action begins with a nightclub performance by the teenage rocker Silver Johnny, whose powerful effect on audiences of young women makes him a valuable property in the developing business of rock 'n roll. In a room above the Soho club, Sweets and Potts, two comedic gangsters who claim to have discovered the young singer, wait for the outcome of offstage contract talks between the nightclub's owner and a promoter. With rock 'n roll music in the background, Sweets and Potts kill time in the manner of Gogo and Didi in *Waiting for Godot*. Joined by Skinny, an employee of the club, and Baby, the owner's son, they play cards. They are still there the next morning when Mickey, another employee, arrives and announces that the club's owner has been found sawed in half and stuffed in rubbish bins. It also appears that Silver Johnny has been kidnapped.

In *Mojo*'s all-male cast of small-time dealers reminiscent of a David Mamet play, the language of power dominates as all the men jockey with one another for control. Mickey assumes the role of boss and issues orders. Skinny performs the role of Mickey's right-hand man, but tries to bolster his status with references to an uncle who was in the RAF. Sweets and Potts display their unity as they voice a chorus of random but ponderous observations. Baby intimidates Skinny, tying him up and threatening him with a gun, and challenges Mickey's position as boss, claiming his prerogative as the owner's son. Mickey chides Baby with knowing nothing about the business. The language of power and attendant posturing masks an essential powerlessness and pervasive fear. Butterworth has described *Mojo*'s characters as 'a bunch of children pretending to be something else' (Tallmer, 2003).

Fear of an unknown external threat motivates the characters to engage in sporadic preparations for defence. They secure the doors, and when Skinny is sent out for sandwiches he uses the money to buy a small gun. The men can only speculate about what it is they fear. The play's violent climax, however, demonstrates that competition within the group creates the real threat. Baby, who seems immune to fear, leaves and subsequently reappears with Silver Johnny, whom he hangs upside down like a side of meat. Baby reports that he stole a car, drove to the promoter's house, killed him, and took Silver Johnny. Under duress, Silver Johnny reveals that Mickey was involved in Ezra's death. With this information, Baby seizes control from Mickey and kills his ally Skinny. Potts and Sweets depart together, as Baby seems to offer Silver Johnny friendship by inviting him to go out for a walk.

The generational conflict at the centre of this play creates a metaphor of post-Thatcher politics. The older generation – offstage and unseen – based its success on exploitation of the young, using Silver Johnny to attract customers and the other young men to do the club's menial labour. They lose their power not through an uprising of the younger generation but through the breakdown of loyalty among themselves. The younger generation, exemplified by Baby and Silver Johnny, gain success not through work but through sex appeal and glamour, on the part of Silver Johnny, or fearless aggression, on the part of Baby. Baby, as his name implies, acts more like a child than an adult. At one point he goes out and buys everyone taffy apples. Even the lethal pistol looks like a toy. When Baby gains control of the club, he immediately validates Mickey's opinion that he lacks knowledge or interest in the business by going out for a walk. Silver Johnny remains passive throughout the violent transition. The club's employees, who have been drawn to Soho by its aura of glamour and their dreams of wealth, provide expendable pawns in games played by the powerful. Only Sweets and Potts, whose oddball loyalty to each other sets them apart, exercise a mysterious independence and remain untouched by the violence.

While the play turns on the accession to power of a younger generation, it does not imply unity among the younger characters. Rather, its action points to the existence of a primal rivalry among men. In their struggle for dominance the men mobilize language, physical force, weapons, and alliances. The rivalry plays out in an environment apparently beyond the control of any external authority and unmediated by morality or law. Such an environment is described by Hobbes in the *Leviathan* as that which prevails in the absence of any communal institution to exert authority or command fealty: 'No arts; no letters; no society; and which is worst of all, continual fear and danger of violent death; and the life of man, solitary, poor, nasty, brutish, and short' (XIII). In presenting this savage struggle for supremacy, with status and financial success as the perhaps illusory goals, Butterworth offers a nightmarish glimpse of the ultimate Thatcherite fantasy, in which there really is no such thing as society.

McDonagh's *The Beauty Queen of Leenane*, a 1996 co-production of Druid Theatre Company of Galway, Ireland, and the Royal Court, shows a generational war of dependence and desire fought between a mother and daughter. It forms the first work in a trilogy that de-romanticizes life in a rural Irish village, emphasizing personal inadequacies, predictable failures, petty acts of vengeance, and a collective narrative that is simultaneously absurd and tragic. The action occurs in a remarkably confined environment, the small and dark main room of a rural cottage in west Ireland where 40-year-old Maureen and her 70-year-old mother, Mag, live. Maureen is a spinster whose previous bid for independence – serving a stint in England as a domestic cleaner – ended in hospitalization for a mental breakdown. Mag is a manipulative and demanding invalid.

Maureen encounters an unexpected opportunity for happiness when an old flame, Pato Dooley, courts her during a visit from England, where he works. Mag, who fears being left alone, uses all the weapons at her command, from shaming to outright deception, to cut off her daughter's escape. When Maureen discovers that Mag has destroyed a letter from Pato, she retaliates with lethal violence, but fails to extract the crucial information that would have allowed her to accept Pato's proposal of marriage. At the end of the play, both women have lost. Alone in the cottage, Maureen listens to a traditional song associated with her lover and starts to rock in her mother's chair, but then deliberately leaves the room, closing the door behind her. This final gesture of ambiguous closure creates an empty space and a moment in the present that resists incorporation into the past or future.

The Beauty Queen of Leenane embeds the generational conflict within the fabric of a village life that paradoxically marries constraint and freedom through its lack of official social authority and its ever-present and all-seeing populace. In such a place, acts like cutting the ears off a brother's dog in a long-running family dispute are talked about but not officially acknowledged or punished. Religion, though an important aspect of social identity, fails to exercise moral authority because the local priest, whose name is consistently mispronounced, is a figure of derision. The life-and-death contest between Maureen and Mag participates in the unchanging landscape of village life, just as its themes have become lodged in the bodies, the daily activities, and even the food eaten by the two women. Ray Dooley, the 20-year-old neighbour who stops by the cottage with messages from his brother Pato and unwittingly aids Mag, seems not to perceive the depth of the division between mother and daughter. His incomprehension, in fact, provides much of the humour in this dark comedy. Ultimately he serves as the blank page on which the history of Mag and Maureen, along with that of the wider village, is recorded. His incompetence as a messenger and observer stands as a reminder of the unreliability of history, especially where there are few witnesses to any event.

This play, in common with other in-yer-face drama, questions the possibility of freedom. Choice is limited by the habits and assumptions that structure the lives of individuals and the entire village. Dreams and ambitions prove futile. All the characters – not only Mag – suffer from disability. Maureen's confidence has been permanently damaged by her earlier mental illness, and her ability to distinguish between reality and illusion seems weak. Pato proves impotent when he spends the night with Maureen, and then must depend on his slow-witted brother Ray to transmit his explanation and attempt to re-establish the relationship. The village culture almost seems to demand disability. In this culture, power is exercised through the language of helplessness, as Mag demonstrates. Both women use childish means to assert will – a pattern of behaviour that further embeds them in the village rather than giving them strength to escape. This isolated and marginal culture, with its habits, assumptions, and faulty history, continues to structure identity and choice in ways that are limiting without being rational.

Questioning of basic assumptions of the feminist movement can be seen in this emblematic rivalry between two intimately connected women. The two women cannot ascribe their oppression to men or find common ground in their sex or gender experience. The men, though they enjoy a greater range of movement than the women, interact with them in ways that do not harm them or subject them to disrespect. The only fault evident in the men is their failure to understand the deep strain of brutal hatred at the basis of Mag and Maureen's oppression of each other. The pattern of each woman justifying her cruelty towards the other by casting herself as victim seems as immutable as the constant rain. The personal, moreover, turns out to have no political resonance in this case. The death of an unpleasant old woman and the private revenge of a troubled spinster cause no change in the life of the village and point to no larger themes beyond the poisoned mother–daughter relationship.

Summary

The in-yer-face plays of the mid-1990s do not speak with a unified voice, but they do show a common impulse to confront audiences with the presence of the post-Thatcher generation. They constitute a decisive break with the previous generation's utopian yearnings, collective spirit, and reliance on dialectical forms of discourse. At the same time, they contest the philosophies or culture of Thatcherism in significant ways. The erosion of individual autonomy by pervasive marketing forms a major aspect of the social criticism of in-yer-face plays. Some plays expose or parody Thatcher's ideal of free market competition by evoking its logical extreme. Many focus on marginalized groups and foster identification with violent, self-destructive, and non-productive individuals, thereby clashing with the middle-class values promoted by Thatcher and the Conservative Party. Situations represent authority as ineffective, irrelevant, or merely absent, rather than a limiting or threatening force. Characters representing the post-Thatcher generation show disdain rather than respect, fear, or defiance towards traditional figures of authority. None of the plays presents a moral viewpoint within the play, as in a character voicing condemnation of certain actions. Rather than signifying freedom, however, this absence intensifies the threats to those who are weak and vulnerable. For those strong enough to pursue their desires, the lack of moral constraint does not increase their pleasure or joy; instead it subjects them to anxiety and confusion. Freedom appears to be an almost meaningless concept because of the difficulty or impossibility of determining what constitutes choice.

A focus on characters and situations that represent coming of age at the end of the Thatcher era draws the in-yer-face plays together in expressing the social and political challenges of a particular historical moment. These plays suggest that the dangers inherent in the transition from childhood to adulthood have become more acute as a result of pervasive exploitation, the breakdown of boundaries, and isolation. They represent young characters as disabled, immature, and unable to make choices or construct meaning. They show the post-Thatcher generation as powerless but unwilling to define themselves as victims. Early death occurs often, but the pervasive mood of loss, and the pain and waste evident in the lives of young characters, overwhelms the emotional effect of any particular death. Though omnipresent, death is usually kept offstage. With no bodies to be taken up, even death loses its defining power. The absence of moral and historical definition negates the potential for tragedy.

The in-yer-face writers seek truth in the portrayal of pain. Their plays focus on pain in a way that cannot be fully explained by describing scenes of brutality and torture. In productions, these scenes have been created with unforgettable vividness and prolonged to the point of great discomfort. Endurance of cruelty, rather than striving for ideals, constitutes the heroic action of these plays. Vicarious participation in suffering serves as a ritual of inclusion for the audience. Ken Urban, writing about cruelty in the popular culture of the late 1990s, argues that the focus on intentionally inflicted pain constitutes 'an exploration of nihilism's ethical possibilities' (355). The plays do not contextualize their immediate and intense, though ephemeral, experience of pain within a structure of values or beliefs, or even within a stable and complete narrative. Pain without such contexts forms the basis of a collective generational consciousness. Collective consciousness does not, however, imply collective loyalty. The alienation evident in the in-yer-face plays extends to a distrust of peers.

Despite the fact that they identify concrete social problems along with a generational crisis of meaning, the in-yer-face plays do not even gesture towards solutions. They locate the rejection of political involvement in a sense of abandonment and exploitation. They emphasize that the conditions fostering their alienation are not addressed by political parties or processes. Rather than calling for change, the in-yer-face plays assert the difficulty of overcoming entrenched habit. Significantly, these plays attribute more of a role to economics than to politics in structuring the lives of members of their generation. With the market now performing the kind of supra-national organizing role once held by religion, economic thinking has become so pervasive that its terminology inhabits individual identity and personal relationships. Material culture forms the basis of recognition, even for the most deprived and isolated of the characters. The playwrights, to various degrees, condemn materialism, identifying it as a source of exploitation, but the characters in their plays fail to free themselves from the habits of materialist consumption. Women who act or serve as representatives of their gender expose gender politics, but the failures of bonding and of collective female power question feminist solutions to the powerlessness evident in many of the female characters. Extreme poverty has displaced class as the marker of social division. Rather than identifying with working-class issues such as wages or working conditions, these plays locate the failings of society in the lives of the truly helpless.

Anger exists in and helps to define these plays, but is as often turned inward as outward. These playwrights seem to be asking for a new politics, but are unable at this point to articulate political concepts that encompass both their specific concerns and wider social issues. They deny traditional forms of agency but do not imagine new definitions or forms. They refuse to externalize their anger, shame, and frustration over these failures by creating enemies to be defeated. Open endings which do not bring closure to the dramatic action signal the unfinished nature of the conceptual processes in which the young writers are engaged and allow them a space for revisiting and revising many of the ideas expressed in this first group of works.

Despite their mood of loss, themes of division, and evocation of social breakdown, in-yer-face plays are known primarily for what Lehmann describes as a 'desperately psyched-up sense of life' (118). The energy these plays brought to the experience of seeing drama reinvigorated British theatre. This energy arises in part from confronting accepted cultural patterns or values, especially those associated with attitudes of political correctness. In-yer-face theatre opposes itself to the type of polemic theatre where an attitude of unanimous agreement is expected. It deliberately provokes disagreement. Walkouts, which break up the sense of audience unity, intensify the excitement of being in the audience and lend a feeling of risk to the experience. Some plays create a mood of exhilaration through the celebration of *hubris* and denial of tragic closure. Rather than tragedy, in-yer-face theatre offers what Howard Barker has termed 'theatre of Catastrophe'. Barker's description comments on the politics and points to the source of energy of in-yer-face theatre:

> Traditional tragedy was a restatement of public morality over the corpse of the transgressing protagonist—thus Brecht saw catharsis as essentially passive. But in a theatre of Catastrophe there is no restoration of certitudes, and in a sense more compelling and less manipulated than in the Epic theatre, it is the audience which is freed into authority. In a culture now so rampantly populist that the cultural distinctions of right and left have evaporated, the public have a right of access to a theatre which is neither brief nor relentlessly uplifting, but which insists on complexity and pain, and the beauty that can only be created from the spectacle of pain. In Catastrophe, whose imaginative ambition exposes the reactionary content in the miserabilism of everyday life, lies the possibility of reconstruction. (54)

In-yer-face theatre was as much an audience phenomenon as it was a playwright one. The in-yer-face plays offered audiences a form of social and even political identity conditioned by events like the Live Aid concerts in which attendance signalled affiliation and constituted a form of collective action. Audiences brought to the theatre a level of intentionality and intensity that gave their attendance the overtones of a political demonstration. By their presence, they validated the stark evocations of cruelty and powerlessness represented by the characters and situations onstage. Simply by attending, audiences performed their openness to the risk at the heart of the dramatic provocations of these plays – the incitement to hope. For, despite their nihilistic aspects, the in-yer-face plays do not present a politics of despair. They exemplify the courage to bear intractable pain, rather than choosing the defeatist and readily accessible alternative of anaesthesia. In the engagement with pity and fear, without the relief of catharsis, resides the hope expressed by Barker that grappling honestly with complexity, pain, limits, and frustrations may initiate a process of political and social reconstruction.

3
Intergenerational Dialogue

The in-yer-face plays announced the potential and power of a younger generation with new questions about British society, assumptions about personal and collective identity, and attitudes towards political action. Their impact reverberated through Britain's closely connected community of playwrights and directors and immediately began to draw responses from established playwrights as well as from a fresh contingent of new playwrights extending and revising themes that had been introduced. The new dramatic phenomenon influenced plays chosen for revival and reinterpretations of classics. The responses, which moved away from confrontation and towards dialogue, have generated further responses from many in-yer-face playwrights as they continue to develop and theatricalize their political perspectives. This dialogue, which confirms the importance of the generational transition while offering viewpoints from both sides of the generational divide, has created the dominant current in post-Thatcher political theatre.

The intergenerational dialogue seems to have captured the political mood of the time, as an older generation looks back at a collective desire for political transformation that remains unfulfilled, a younger generation looks ahead to a world that has been transformed in ways not envisioned by the previous one, and both generations explore their power in relation to each other. The themes of this dialogue, laid out in the first in-yer-face plays, have revolved around history, identity, freedom, and action as expressed in relationships that represent the political configurations of contemporary Britain. Participants include a wide spectrum of playwrights using the parameters of the dialogue to explore salient political issues.

The generational dialogue in recent plays by Caryl Churchill

Caryl Churchill, whose encouragement and acknowledgement of young playwrights has been evident throughout her long involvement with the Royal Court Theatre, engaged in dialogue with the work of the in-yer-face playwrights from its inception. Churchill had often placed political and economic issues in a generational framework, notably in such plays as *Cloud Nine* (1979), *Top Girls* (1982), and *Fen* (1983). Her mid-1990s plays, *The Skriker* (1994), and a translation of Seneca's *Thyestes* (1994) anticipate themes of the in-yer-face plays by highlighting the exploitation and destruction of the young. In her plays of the late 1990s, Churchill extended and further interpreted the generational theme of the in-yer-face plays and her own mid-1990s work by initiating an intergenerational dialogue. These plays make visible issues of identity and history by showing members of the older and younger generations struggling to comprehend and be understood by one another. They highlight the intense need of parents and children for each other, both in the private realm of identity formation and in the public realm of national continuity, history, and legacy. Although they acknowledge difficulty and warn of even greater perils in the future, they evoke the potential for human relatedness even in the midst of disaster and find hope for future societies in the human capacity for connection inherent in the kinship between generations.

Blue Heart brings fresh perspectives to its exploration of intergenerational dynamics and issues of identity and continuity in the pair of related one-acts *Heart's Desire* and *Blue Kettle*. Co-produced by Out of Joint, the Royal Court, and the Theatre Royal Bury St. Edmonds, *Blue Heart* was first presented at the Traverse Theatre, Edinburgh, in 1997. Both one-acts focus on interactions between parents and children, and the unexpected issues and outcomes of these interactions. *Heart's Desire* shows a trio of middle-aged, middle-class people preparing lunch, chatting, and scolding an intoxicated son when he intrudes. They wait eagerly for the arrival of Alice and Brian's daughter, who is returning from Australia for a visit. Anticipation of her homecoming exposes both habitual patterns and deep desires and fears. The everyday setting and realistic situation become absurd as time after time the action stops, then begins again at a previous moment and lurches forward. Words and phrases mutate as they are repeated, speeding up, slowing down, and becoming telegraphic or disjointed. The interruptions seem to occur at random, but replayed actions often bring different responses. The couple's deep animosity towards each other plays out in different ways, as Alice variously leaves the marriage, confesses to a long-term affair, and demands that Brian leave.

Against this repeating set of habitual actions Churchill juxtaposes the potential for the unexpected in frankly theatrical moments that occur each time the doorbell rings. With each ring the opened door admits an improbable intrusion, from a pair of armed assassins to a group of lively children, a uniformed official, an enormous bird, a young Australian woman who introduces herself as their daughter's lover. Each intrusion briefly takes over the action, but as soon as the intruders exit, the same patterns resume. The daughter's arrival, rather than producing closure, brings only another stop in the action and another start from near the beginning, before the final blackout.

The warped and often hilarious action of *Heart's Desire* performs the instability of narrative and the impossibility of history. In Pirandellian fashion, each character brings a starting point to the story, but the story itself keeps changing. None of the many versions made available in the course of the play stands as the authoritative one. What does get communicated through the confusion of starts and stops and changing viewpoint is the suffering of individuals in this family. Each character inhabits a microcosm of profound pain. Alice feels isolated in a loveless relationship. Brian voices an agonized and ambivalent desire to eat himself that exposes his self-loathing and longing for self-possession. Maisie, the oddball aunt who offers bits of meaningless conversation to bridge the silences brought on by the conflicts between Brian and Alice, suffers from an acute fear of dying. The son remains marginalized in the family home. Like Pirandello's characters looking for an author, all the members of the family expect the absent daughter to understand and define their situation when she returns. Like Beckett's characters, they fill their endless present with repetitive words and habitual actions while they wait for the defining encounter.

In this family Churchill presents a metaphor for a dysfunctional nation divided by deep and long-standing conflicts, suffering in diverse ways that can scarcely be described, and mired in repetitive patterns. Communication has broken down almost completely in the family, and only the inertia of habit sustains it as a collective entity. While ostensibly living together, each member exists in a separate sphere of pain. They depend on the arrival of the daughter, who is part of the family but separate from it because of her residence abroad, to give their existence meaning. The family members are too preoccupied with their personal stake in the ongoing battle to consider what they are demanding of their daughter, and ultimately her arrival makes no difference to the family dynamics. Brian and Alice do not move towards the door, but remain fixed in their places. When she enters, Brian attempts to express the immensity of his feeling for her, but only begins, 'You are my heart's—' before the action breaks off (36).

In *Heart's Desire* Churchill playfully turns the table on the in-yer-face dramatists while seconding one of their major arguments. She shows a younger generation that has deserted the field of battle rather than face the previous generation's need and expectation that it will be the final arbiter of their conflicts. She balances this portrayal against that of an older generation too self-involved and stymied by habitual conflicts to offer realistic expectations to their offspring. The conflicts that remain unresolved and the actions that remain uncompleted point to the impossibility of a collective identity or history. The play does, however, contain a strong element of hope in the fact that each time the doorbell rings, someone opens the door.

Blue Kettle, the second play in the *Blue Heart* pair, offers a view of false identities and histories in the context of exploitation. It focuses on 40-year-old Derek, an unemployed man who contacts a series of elderly women, telling each one that he is the son she gave up for adoption many years previously. The first scene, when Derek meets Mrs Plant and she describes the circumstances of his conception, seems to initiate a conventional realistic drama of discovery, connection, and meaning. Subsequent scenes shift the style from realism to black comedy, as Derek acts the part of long-lost son in reunion dramas with four more women. In scenes with his girlfriend and actual mother he claims that he plans to get money from these women, but he does not carry out this intent. The confusion of identities and stories increases when Derek and his girlfriend have dinner with one woman and her husband, and later when he brings two of the women together to meet. Even when confronted with evidence of Derek's deception, the women want to continue believing the original story. Questioning and rejection of Derek's identity and narrative occur only at the end when he changes his first story, telling Mrs Plant that he knew her son and was with him when he died during service in the army.

Churchill intensifies this play's absurdist style by beginning, in the second scene, to replace random words of dialogue with the words *blue* and *kettle*. As the scenes progress, these word substitutions occur more often, and then they begin to disintegrate, losing syllables. By the final scene, the characters utter sequences of single letters. The play thus comments not only on the elusiveness of truth in life narratives and claims of identity, but also on the instability of language. Language, the primary means of definition, changes as it is used, frustrating memory and thus identity and history. The scene between Derek and his actual mother, who suffers from dementia, further emphasizes the fragility of memory.

Against these challenges to communication and empathy Churchill places what seems to be a strong and even desperate drive to connect. True, this drive does not exist in every character: Miss Clarence, the frosty Oxford don, displays a complete lack of interest in the son she bore. Most of the women, however, persist in believing the story that established the connection, even in the face of contradictory statements. They welcome the relationship and seek to continue it, even though they have families. Derek himself gets so involved with the women he contacts that his girlfriend describes him as obsessed. Churchill thus counters plays like *Mojo* and *Shopping and Fucking* that depict humans as incapable of sustaining relationships, with a reminder that the human need for relatedness will contend with the most difficult obstacles to establish and maintain connection.

This play's political metaphor centres on the man who claims to have been abandoned. The story of a relationship severed when an infant is given up for adoption but later re-established has long occupied a place in the popular imagination, from ancient Greek tragedy to current daytime television. Yet the story of abandonment in this particular case is false. Instead, what turns out to be demonstrably evident is that the man who claims to have been given away finds a surplus of potential mothers in the form of women who wholeheartedly accept him as their son. They welcome the connection and its related narrative of completion in which their lives have followed a meaningful trajectory ending in wholeness. In this pair of short plays Churchill offers a glimpse of what the generations want from each other and how each one uses the other. The older generation depends on the one that comes after it to write their history and thus assign significance to the conflicts and pain that have been central to their lives. When the younger generation inherits the material of the past, they gain power to use, distort, and even fabricate parts of it to advance their own interests or support their own preoccupations. Tensions arising from a basic conflict of interest characterize the relationship, as the older generation seeks completion of itself in the younger, while the younger generation tries to take what it needs to build something new. In spite of the tensions, the generations remain in contact, in negotiation, tied together by the distinctive and compelling habits of family, culture, or nation.

Churchill's *Far Away*, first presented at the Royal Court in 2000, continues the generational dialogue in a serious and perceptive treatment of questions about the potential for individual or collective action in an environment where conflicts and threats multiply as the potential for harm escalates. This play focuses on Joan, a young woman who comes of age amid unstable systems of meaning and intensifying violence. Its three segments move stylistically from realism to expressionism to absurdism. In the first scene, Joan is a child who has been sent to visit relatives on an isolated farm. Sleepless, she has come to Harper, her aunt, for comfort. Played by an actual child, dressed in a nightgown and holding a stuffed toy, Joan stands out as a figure of vulnerability and innocence. At first she seems to have been awakened by country noises like the cry of an owl, and Harper gives reassurance. Joan, however, goes on to reveal that she has gone outside, slipped in blood on the ground, and observed her uncle forcing people out of a lorry and into a shed. Harper continues to supply ordinary reasons for these things – a 'little party' (8), a dog run over in the road – and coaxes Joan to forget about what she has seen and go back to sleep. Joan nevertheless persists in questioning, offering further details of what she observed until the full picture emerges: the people in the shed had been beaten, and her uncle was hitting a man and a child with a rod. Harper reluctantly acknowledges what Joan has witnessed and offers to 'trust' her with 'the truth': 'Your uncle is helping these people. He's helping them escape' (12). Awkwardly but insistently she explains the injuries and beatings as the work of others and the actions of traitors. Finally, she invites Joan to participate in the 'big movement' to 'make things better' (14) by helping clean up the shed in the morning.

Without further development of the action introduced in the first scene, the play moves ahead to Joan as a young adult. The second segment, in several scenes, shows Joan taking up her work as a designer of hats, shaping felt, and applying embellishments to produce creations that parody the displays of fashionable millinery by the wealthy spectators at Ascot. As they work, Joan and her workmate Todd chat about the parades in which the hats are judged, the trials constantly being broadcast on television, and the corrupt management of their industry. The parade, which climaxes the second segment, shows wretched ranks of chained adults and children being marched across the stage wearing elaborate hats including the two made in previous scenes. In its aftermath Todd congratulates Joan on winning the hat competition. Joan expresses sadness that most of the hats were burned 'with the bodies' of the prisoners who modelled them (25).

The final segment moves ahead another few years and returns to the setting of Harper's farm, opening with what seems an ordinary incident as Harper and Todd discuss exterminating wasps. Almost immediately, however, their conversation reveals the bizarre extremes of a world in which every element is at war with every other element. Different species of animals, from wasps to elephants, along with people in different national and occupational groups, such as engineers, dentists, and car salesmen, have allied themselves with various sides in a global war that has already recruited such natural phenomena as the weather and threatens to make use of gravity. Joan and Todd are married but separated by their roles in the war. In the final scene they have both found their way back to Harper's farm for a brief reunion. Joan is exhausted and sleeping. Harper suspiciously questions Todd about his loyalties and scolds Joan, when she wakes up, for exposing her to danger by coming there. Ignoring Harper, Joan and Todd embrace. Then Joan describes the obstacles she overcame in order to experience this moment of peace: the thunderstorms, rats, and Chilean soldiers, the heaped up bodies of people killed by 'coffee . . . pins . . . heroin . . . petrol . . . hairspray' (37–38), and finally an enigmatic river.

This play traces the breakdown of idealism and the subsequent deterioration of human society to the point where the world is governed only by the fight for survival. The first scene shocks with its depiction of violence, not because the violence is explicitly represented but because of the child's horrified reaction to what she has seen. Harper's cynical use of the rhetoric of idealism to justify the cruelty witnessed by Joan stands in ironic contrast to Joan's innocent trust that people, especially children, will be treated kindly. With disturbing ease Harper undermines the strength of Joan's simple logic and overcomes her straightforward moral judgments by convincing her that the suffering people she has seen are being helped. Joan is alone, with no ally or alternate authority who might support her own perceptions. She therefore accepts Harper's weakly fabricated version of reality and agrees to join an endeavour which Harper promises will bring spiritual rewards. As an apologist for tyranny, Harper does not participate directly in acts of violence and oppression, but facilitates them and ultimately corrupts her young niece.

The second section serves as an analogy for the current situation. At first it seems to take the form of social drama with the quirky, fantastical touch of hat-making as a state-run industry. Joan's sense of accomplishment in mastering an art form and Todd's determination to seek improved working conditions follow the patterns of narratives about self-realization and social progress through effort and rational dialogue.

Casual references to trials, however, destabilize this sense of familiarity and lead to uncertainty. As Elaine Aston (*Feminist Views*, 2003) points out, Joan has learned from her childhood experience not to ask questions or look too closely at what is going on around her, but the references prompt audience questioning. The parade of condemned prisoners wearing the completed hats brings horrified realization. The shocking contrast between the chained and ravaged bodies and the extravagant millinery intensifies the Kafkaesque brutality of the first scene, in which beatings were described by Joan as 'help'. Todd's ideals of honesty and fairness may still be tolerated, but they are no more than distractions from the mass trials and executions carried out by unseen authorities. Artistic endeavour may still be encouraged and rewarded, but its products serve only as obscenely inappropriate adornments for those condemned to die.

In the final section of *Far Away* Churchill goes even farther than the in-yer-face dramatists in exploring what Urban has identified as 'nihilism's ethical possibilities' (355). The nightmarish world of this scene represents an extreme of social breakdown that could justify nostalgia for the lies and wilful blindness of the earlier scenes. Traditional values and expectations have all but disappeared, evident only in Harper's condemnation of mallards for committing rape. Their absence leaves individuals with no compass as they attempt to navigate through the grotesquely disorienting morass of deadly conflicts surrounding them. Threats are everywhere, and nothing with any form of power can be trusted, because loyalties shift continuously and without warning. While not only humans but also animals and the elements of nature engage in killing one another, there is no sense of what any of them might be fighting for, other than individual survival. Even the generational contract has broken down, with Harper showing only anxiety over Joan's visit.

Joan, in an act of heroism, breaks with the norm of suspicion and hate that seems to define everything around her. In the midst of this chaos of killing and destruction, she has abandoned her combat role and risked everything for even a brief respite with the person she loves. She thus demonstrates the power of love to motivate an existence based on more than the will to survive. To reach Todd she has overcome not only the tangible dangers around her but also her fear of uncertainty, which she confronted when she reached the river. As she describes the journey Joan says that she hesitated because she had no way of knowing what the river would do to her: 'it might help me swim or it might drown me' (38). Since she had to cross this powerful barrier to reach Todd, she finally submitted to uncertainty and stepped into the water. Joan's love demanded from her this act of courage, which involved an alliance rather than a confrontation. Though based on personal desire and commitment rather than abstract ideals, her action stands as a sign of hope.

With *Far Away* Churchill challenges the in-yer-face playwrights' arguments against activism through Joan's example. Joan grew up being lied to and came of age in an authoritarian state that used the products of her talent in lurid displays of its power and cruelty. Plunged into an Armageddon, forced to confront devastation while fighting off threats from nature and manufactured substances, Joan nevertheless finds in her love for Todd a compass for her extraordinary journey. Acknowledging the limits of idealism, the instability of narrative, and the unknowability of reality, Churchill allies her viewpoint with that of Sarah Kane in suggesting that loving another creates the basis for meaning even in the extremes of chaos and threat. Though this play offers only the barest sketch of a route for its Brechtian way out of moral and environmental catastrophe, it argues against defeatism. Through its imaginative and convention-breaking form the play, in common with much of Churchill's work, also encourages unconventional ways of viewing the world, politics, and other people.

A Number (2002) considers issues of identity and power through the device of an encounter between an aging father and three genetically identical sons. The first father–son scene establishes that the son Bernard is one of a group of clones created from an earlier Bernard. At age 35, Bernard has just discovered that he is a clone and that a number of others with the same genetics are living. He has come to his father for an explanation. The father Salter, like Harper in *Far Away*, concocts a comforting story to mask the harsh truth. In subsequent scenes the truth emerges in spite of his efforts. Salter had his first son, conceived with his wife, taken into care, because he could not handle him after his wife's suicide and while struggling with disabling alcoholism. Later, when he had recovered, Salter had the first Bernard cloned so as to make a new start with 'one just the same because that seemed to me the most perfect' (14).

In their separate scenes with Salter, the sons reveal significant differences. The 35-year-old Bernard, mild-mannered and self-contained, shares an uncomplicated closeness with Salter and defines him as a good father. The original Bernard, whose appearance and speech mark his marginal social position, brims over with despairing hatred of both his father and himself, and communicates violence in every word and gesture. After the two sons meet offstage, the second Bernard goes away, seeking a place of safety from the brother who frightens him and distance from the father he has begun to hate. His attempted escape, however, fails. The older Bernard kills him, achieving revenge against Salter, and then commits suicide. Having lost both Bernards, Salter finally meets with one of the individuals cloned from the original Bernard's cells, still seeking a son. This young man, Michael Black, demonstrates a warm and relaxed manner not seen in either Bernard. His life, as he describes it, contains all the elements of unremarkable happiness. He loves his wife and children, enjoys his work, and finds a sense of belonging not only in his genetic sameness to the other clones, but also in his genetic similarities to other living organisms, from chimpanzees to lettuce. Michael Black's affirmation of life does not console Salter. He still wants a son.

A Number explores the desire and power at the basis of the relationship between generations. The sons want love and truth from their father. While these desires seem simple, they form the basis of the complex process through which a human knows and values the self. As it happens, this father has, in a sense, given one son love and the other truth. The younger Bernard has loved and been loved in tangible and demonstrative ways by his father, but has not been told the truth about his origin or connections. Even when he confronts his father, Salter avoids telling him the complete story of the events that led to his birth. When he learns from his brother the full extent of his father's lies, the younger Bernard flees the relationship. The older Bernard cannot value himself because he has not experienced love from his father. Even before he was sent away, the first Bernard felt abandoned, because his father did not come to him when he cried out at night. Although Salter insists that he did love him, the first Bernard has no memory of love and therefore no starting point from which to love or value others. When he realizes that his father not only is capable of love but actually has given love to the substitute for himself, his rage destroys both him and his brother.

The father's desires are both simple and complicated. On a basic level, Salter needs companionship as he ages. He prefers the second Bernard, but is willing to resume a relationship with the older son even after he has killed his brother. Salter's desire goes beyond companionship, however, as he seeks in his son a definition and completion of his own existence. In starting over with the cloned second son, Salter seeks a kind of earthly salvation – a second chance to give his life a positive definition. This son's successful maturation allows him to feel that he has atoned for the wrong he did with the first and achieved new definition. As he says, 'I did some bad things. I deserve to suffer. I did some better things. I'd like recognition' (35). The new definition, however, depends on the second son not knowing the truth; revelation of what Salter has done ends the relationship. When Salter loses both sons, he also loses the possibility of completion in his own life. His interview with Michael Black does not fill the void and leaves Salter so visibly depressed that Michael offers a brief apology for his own happiness.

As the father, Salter holds a traditional form of power as the progenitor, which is extended in the play's scientific conceit centred on human cloning. The older Bernard refers to his father as a 'dark dark power' (15). Salter has assumed god-like powers of creation and redemption that stand in strange contrast to the self-image he projects in his narrative of helpless addiction and depression. In using this power to save himself Salter becomes a monster of egotism, with his sons as mere extensions of his ego. Salter discovers the limits of his power when the son who has been relegated to non-existence reappears. When the returned son murders his brother and kills himself, he puts an end to any possibility that Salter may reassert his ego through the younger generation. Salter, however, does not fully realize the limits of his power until he meets Michael Black. In spite of his identical appearance to the two Bernards, Michael brings with him a separate history in which Salter has no part. Michael demonstrates that history, not genetics, forms the basis of family.

Though his time on stage is brief, Michael Black plays a crucial role in the theme, rather than being merely an amusing coda to this drama. He frees the play from the traditional opposition between the good son and the bad son. By understanding and accepting relatedness with even such humble organisms as lettuce, instead of asserting an egotistical uniqueness, Michael presents an alternative to traditional masculine and Western subjectivity. His love reaches out in many directions, rather than being channelled into one type of possessive relationship. He ends Salter's dominance, not by opposing his position or power but by introducing an entirely new perspective. Michael, no less than the two genetic brothers who were parented by Salter, comes from a family, though one not present in the play. It remains the enigmatic alternative to the egotism, conflict, and destruction that is represented.

With a dysfunctional family once again serving as metaphor for a nation in conflict, *A Number* probes the mechanisms by which an older generation has both abandoned and exploited its young. Through failures of love and by withholding truth in an egotistical attempt to maintain power, the parents have generated conflicts among the young and set them on a path of rage and destructiveness that threatens the future. The antidote to this conflict, suggests Churchill, is not a war between generations, but alternate forms of identification. A world that can, in the foreseeable future, expect to see human cloning become a reality should be able to create paradigms for new subjectivities, forms of relatedness, and histories that might lead to the salvation that eluded Salter.

Churchill's contributions to the generational dialogue offer thoughtful, if tentative, hope for addressing the dilemmas introduced by the in-yer-face plays. Her recent plays metaphorically invoke a divided nation, mired in the type of conflict associated with the earliest of human narratives, yet treading on new ground in human experience. Images of chaos, destruction, and isolation juxtaposed with images of the uncorrupted young emphasize both the possibilities and the danger in a world that is mapping new frontiers of scientific and political power without moving beyond age-old patterns of behaviour. These destructive patterns include lying and manipulation, using ideals to mask egotism, tyrannizing the weak, and failing to take responsibility for one's actions. The instability of language and narrative seen in these plays indicates the difficulty of establishing identity, communicating with others, and building a common base of experience that can be expressed as a collective narrative or history. These difficulties have not, however, annihilated the human capacity to love, which can enable an ordinary person to overcome extraordinary obstacles. With Sarah Kane, Churchill finds the impulse for a better world in love that puts the welfare of another before that of oneself. She goes beyond Kane, however, and returns to some of her own early preoccupations, to suggest the possibility of a non-objectifying love transcending ego and reaching out to all being. Supporting Ravenhill's conclusion in *Some Explicit Polaroids*, she does not envision hatred or oppression being overcome through confrontation or opposition, which only perpetuate patterns of aggression, but rather through the strength and completeness of an alternative subjectivity and the non-possessive love potentiated by such a subjectivity.

Revising the terms: Racial and ethnic identity in generational context

Instead of replying directly to the in-yer-face dramatists, a number of new playwrights joined the generational debate by revising its terms. The in-yer-face plays, by ignoring issues of race, region, class, or ethnicity, mirrored the view of British society implicit in the refusal of the Conservatives to address such issues, despite race riots, strikes, and growing economic divisions during the Thatcher years. Younger playwrights whose political commitments arise from their experiences as members of a minority race or culture have placed issues of race, ethnicity, and class at the centre of the generational divide that metaphorically represents the British nation. Their plays make visible these sources of identity and conflict by situating questions of history, identity, and choice within specific social or geographic communities. They reinstate the materialist approaches of social realism, historicism, and dialectic as they explore the conditions that create similarity and difference, division and unity in post-Thatcher Britain. While they do not resurrect the idealistic hopes of the 1970s social movements, they return to issues made visible by those movements. They emphasize relatedness rather than isolation and momentum towards the future rather than imprisonment in the present.

Expressions of cultural identity within immigrant communities, such as the Notting Hill Carnival, began in the early 1960s, but theatre based in the identities and experiences of non-white British people has been slow to develop. Currently, only four companies focus on the experience of ethnicity in Britain. Tara Arts, founded in 1977, focuses on the cultural heritage of British Asians. Talawa Theatre Company, which began producing plays in 1986, develops new writing by British dramatists of West Indian and African descent. Tamasha Theatre, founded in 1989, has sought out new plays based in the experiences of Asians in contemporary Britain. Nitro Theatre Company (formerly the Black Theatre Cooperative) mounts occasional music theatre productions written by and featuring black artists. None of these companies operates its own venue, but their productions have been crucial in nurturing the development of black and Asian theatre artists. The Theatre Royal Stratford East, the Tricycle, and the Hackney Empire, in addition, actively address non-white audiences. The Royal Court has introduced such playwrights as Mustapha Matura, Hanif Kureishi, and Winsome Pinnock, and the National has premiered recent work by playwrights of colour.

Representation of ethnic identities within mainstream culture entails negotiation with complex political problems. British society includes individuals of colour, Asian and African cultural communities, and unresolved tensions arising from racism and ethnocentrism. These individuals, communities, influences, and problems constitute important elements of contemporary British politics. At the same time, the history of misrepresentation and objectification has created resistance to identification based primarily on national origin or skin colour. Mixed backgrounds often complicate artists' identities, and differences within groups perceived as Asian or Black create another source of controversy. Constructing collective identity is a political act. Such an act contains the potential for objectification, and does not invariably lead to empowerment of minority individuals and groups. Nevertheless, the exploration of ethnic and racial identities provides an indispensable perspective on national political life.

The change of identity from one generation to the next in families who have immigrated to Britain provides a rich context for exploring history, identity, and choice. For families in which an older generation was born outside Britain, the differences between generations may be marked by divergence and conflict. Individuals with ties to more than one geographical or cultural home exemplify the unstable identities defined as postmodern in a very concrete way. Immigrant families always negotiate between two or more systems of meaning and patterns of action that may be inconsistent with each other. Identity and choice must be positioned within competing sets of expectations. Reflections on the immigrant experience not only provide a window into the complexities of that experience but also encapsulate and make visible the negotiations involved in the ongoing construction of a multicultural national identity.

Roy Williams made his first appearance in 1996, with a play he had written while a student. That play, *The No Boys Cricket Club*, was produced by the Theatre Royal Stratford East, which has developed a strong Afro-Caribbean constituency in the community surrounding the theatre. Williams's plays have since reached a broad range of audiences in theatres as well as on BBC radio and television. He focuses on the interplay of race and ethnicity with issues of class, the youth subculture, spirituality, and sexual politics. His plays explore power on multiple personal and social levels. They articulate pride in his ethnic heritage and an ethical framework for action while demonstrating an understanding of human weakness and failure.

Abi Carter, a 60-year-old widowed woman who lives in London with her two young adult children, is the central character of *The No Boys Cricket Club*. Worried about a son who deals drugs and a daughter involved in gang fighting, Abi has lost the optimism and energy that brought her from Jamaica to England. When her best friend from childhood, whose son was killed in racist violence, declares that they should return to Jamaica, Abi confronts her past choices and considers the future. Episodes in the style of magical realism return Abi and her friend Masie to Jamaica and their youthful selves. The imaginary but vividly theatricalized visit reunites Abi with the sense of belonging she lost when she left Jamaica and rejuvenates the pride she felt when she was the star batter for an all-girl cricket team. It also forces her to admit that she sacrificed her youthful dreams for the sake of others. Ultimately, Abi chooses to remain in London but returns to her present reality determined to make changes. Her final moment of connection with her youthful self suggests that she has indeed found a source of strength.

This play gives visibility to the pre-Thatcher immigrant generation, which broke with familiar identifications and initiated new ones. The realistically created Abi and Masie coped with poverty and abandonment, relying on each other because their parents were dead or absent. Though their environment offered few opportunities, the two found an activity that enabled them to gain strength and confidence, display skill, and nourish hope for future accomplishment. For a time the cricket team was the centre of the girls' lives, but it broke apart after a crucial defeat by an all-boy team just as they recognized their imminent womanhood. Marriage and motherhood, which followed, demanded the abandonment of their personal interests and pleasures, and the two have been living for their children. The death of Masie's son and the risky behaviour of Abi's children question that choice and propel the women's journey into their own history. They rediscover the past for their children as well as themselves. Ignorant of family history and Jamaican culture, the children have defied norms of family respect central to Jamaican life. Abi's daughter Danni has refused to eat at home, and her son Michael struck her when she flushed his drug cache down the toilet. Disrespect towards the mother they perceive as powerless has increased the difficulty Michael and Danni have in valuing themselves.

In the Jamaica episodes Abi rediscovers belief in herself and hope for the future. When she meets her younger self, the girl does most of the talking. Young Abi cautions her not to romanticize life in Jamaica, reminding her of the guilt she felt when her mother died and the pain she endured when her father abandoned her. When young Abi bowls and mature Abi demonstrates undiminished batting skill, she realizes that she still has the power to accomplish things. Abi uses this renewed sense of personal power to re-establish authority as a parent. Danni questions her mother's capacity for perseverance, but in reply Abi suggests a game of cricket and, in the final moment of the play, brings young Abi into the London scene for the first time. The style of magical realism that makes visible the imaginary trip to Jamaica and its subsequent effects suggests that magic and mythologizing are inseparable elements of history and basic to its power to influence the future.

The immigrant experience frames the generational transition in this play. Abi has ties to a homeland unknown to her British-born children. Coming to England was one choice that has since dictated others. Struggling with a strange culture and limited in her employment opportunities, she has been unable to communicate the experience of personal authority and choice. The children, who have grown up seeing their parents as outsiders and low-wage workers, do not understand the hopes that impelled them to emigrate, yet are expected to vindicate their parents' choice. Coping with disappointments and limitations, the parents themselves have lost touch with their original dreams and are powerless to deal with angry and disillusioned children who have sought status and belonging in gangs and drug dealing. The play's antidote to hopelessness and anger lies in the realm of art, with stories and recollections that connect children with the dreams and strengths of their parents.

In his subsequent work Williams continues to contextualize social change within the lives of British Afro-Caribbeans. *Lift Off*, produced in 1999 through collaboration between the Royal Court and the National Theatre Studio, focuses on the post-Thatcher generation. Highlighting the perception of blackness as a model of masculine identity and status, it presents two boys who differ racially but both want to be black. In contrast to the metaphorical playgrounds of early in-yer-face drama, the setting is a realistic playground on a West London council estate. Mal, who is black, and Tone, who is white, live on the estate and use the playground, in the manner of adolescents, as a zone of experiment with identity, power, and choice. For both boys, the Jamaican gangsters of their area offer a model of black maleness, status, and freedom to which they aspire. They believe they must be hard – unmoved by danger or suffering – to be accepted by a gang. Both boys boast about sex, fights, and crime, to demonstrate hardness and claim gang connections. Mal excels in hardness, while Tone attempts to imitate his speech and action. Tone draws Mal's derision when he stumbles over the slang term 'irie man', but after winning an arm-wrestling bout Tone boasts, 'I'm blacker than yu, Mal' (12). Both Mal and Tone prove their hardness by bullying a younger black neighbour, Rich, who plays with paper airplanes and resolutely avoids fighting.

Scenes alternating between present and past, with the double-cast Mal and Tone appearing as adolescents and men in their twenties, show the pair blending friendship and competitiveness. In young adulthood Mal still serves as leader and role model for Tone, who envies his friend's constantly ringing mobile phone, criminal exploits, and sex life. Mal, however, has leukaemia and little hope of a cure, since the register of black bone-marrow donors is too small to provide a match for him. Still adhering to the code of hardness, he hides his sickness. A rupture with Tone, caused by his treatment of Tone's sister and the interference of Tone's racist girlfriend, gets Mal thinking about Rich, who committed suicide. Mal visits this dead friend in imagination and finds some relief for his depression and fear. In the final scene, despite their differences, Mal and Tone resume their friendship and its long-established pattern in which Tone follows Mal's lead.

In *Lift Off* Williams examines race relations through the lives of urban youths who not only accept black culture, but also admire and emulate a particular black stereotype. The young men's ideal, which accords with Thatcherism in some respects, combines toughness and self-sufficiency. Mal achieves his status on the basis of his tough-guy, go-it-alone image. In spite of an episode in which his pugnacious assertiveness gets him beaten up, Mal rejects the obligations of relatedness. His disease, however, along with the fact that the only possible cure lies in a donation of body tissue from a racially similar person, teaches him the value of relatedness. Communing with the dead Rich awakens Mal to his repressed love for the peace-loving child and suggests unexplored choices. Tone learns that his need for closeness with the familiar and powerful Mal is stronger than the family loyalty that urges him to avenge the wrong done to his sister. In their own ways both young men negotiate the complex race relations of contemporary Britain, where racial differences may be overridden or intensified by ties of family, neighbourhood, class, and generation. Their rapprochement at the end of the play indicates the importance of connection with one's peers. This conclusion historicizes race relations, presenting race as a changing construct and the relations between racially different individuals as a product of evolving social factors.

Ayub Khan-Din's *East Is East*, produced by Tamasha Theatre Company, the Birmingham Rep, and the Royal Court, and first performed in 1996, played an important role in articulating tensions around the issue of ethnic identity within Britain. Khan-Din, who grew up in an ethnically mixed working-class family in Salford, an industrial area in greater Manchester, has described the play as autobiographical (Olden, 1999). Set in a fish-and-chips shop and the adjoining family home, the play realistically portrays conflict between two generations of the Khan family. George Khan, a Pakistani Muslim immigrant, expects absolute obedience and unquestioning respect from his wife and children – a style of family interaction that he associates with his culture of origin. He demands exemplary behaviour from them as a means of countering the contempt for Pakistanis he has endured during his 30 years in England. When the children, in their teens and early twenties, fail to meet their father's expectations, he grows increasingly frustrated and blames these failures on his white, British-born wife Ella while idealizing the wife left behind in Pakistan.

Divisions in this family encompass both material and symbolic issues, but symbolic issues create the family's decisive crisis. George has discovered that his youngest son has not been circumcised, as required of Muslim males, and he demands that 12-year-old Sajit have the operation. He has initiated marriage negotiations for his older sons Abdul and Tariq. While the five sons and daughter pretend compliance, they mock their father behind his back, calling him Genghis, and do their best to evade his rules. Twenty-one-year-old Tariq sneaks out at night, and eighteen-year-old Saleem is studying art rather than engineering at college. Sajit retreats into a world of his own inside the parka he wears constantly. The mother, Ella, mediates between her children and her husband while marshalling the children to help in the shop. When forced to choose, however, she takes the children's side, and endures George's beating. In the end, the father's determination to arrange their marriages forces the sons to confront him. When he attacks their mother, the children cooperate in defending her, discovering in this action a new unity and definition of family.

The play revolves around the negotiation of identity, in which all the members of the family participate. Based on his understanding of the culture he left behind 30 years previously, George believes he should have the power to define each family member. However, the family lives in a Western society that considers the individual, rather than the family, the basis of identity. Despite what their father may attempt to teach them about loyalty to cultural traditions, they have been influenced by his example of separation from his original culture and their mother's example of defying her own culture to marry a man outside it. Each member of the family has developed individual traits and desires that are incompatible with the power George considers his by right. When Abdul complains to his mother that George makes decisions about his life 'as if he owned it', Ella replies, 'He thinks he does' (16). His inability to relate to them on their own terms makes George an outsider within his family. He speaks broken English, while Ella and the children speak fluently. They, in turn, do not understand Urdu or Arabic. He takes an avid interest in events occurring in Pakistan, while the rest of the family dismisses these developments as irrelevant. Although he acts the tyrant with his wife and children, George vents his despair over the inexorable loss of connection with his original culture in bouts of sobbing when he thinks he is alone.

Like the fathers in Churchill's plays, George seeks completion of his identity in the next generation. He needs the respect and obedience of his children not only to salve the wounds inflicted by a racist society but also to validate Pakistani traditions within a society that treats him and his culture as inferior. George needs his children to exhibit Pakistani identity, because they are his only way of maintaining that identity for himself. His strong investment in an attempt to create Pakistani identities for them is evident in their names, which contrast with George's own English name. The children do not see themselves as extensions of their father, but only Tariq contemplates leaving the family, as did their eldest brother who became a hairdresser. None of the younger family members knows quite how to describe their public identity, as they show by making different suggestions – Anglo-Indian, Eurasian, Paki, English – in one scene. Nevertheless, in the course of the play each negotiates a separate identity within the context of the family. Abdul, taking a cue from his mother's practicality, becomes the worker indispensable to the family business. Maneer finds an identity in religion, though he insists that his father should not force Islam on his siblings. Saleem develops his talent as an artist and rebelliously takes female genitals as a subject for one project. In conversations about Sajit's circumcision, the daughter Meenah demonstrates a familiarity with male anatomy incompatible with the strict purity her father expects. Sajit, whose nervous habits have earned him the family nickname Twitch, attempts to wall himself off from his family by retreating into his parka and shutting himself in the coal shed.

The much-anticipated meeting with the man whose two daughters are the prospective wives for Abdul and Tariq reveals the extent to which George is prepared to compromise his beliefs about family and tradition. After formal Islamic greetings the visitor criticizes Meenah's wearing of a sari rather than the salwar kameez associated with Pakistan. He also disapproves of Ella's smoking, but seems satisfied that the prospective bridegrooms have been brought up to respect tradition. He does not wish his daughters to live in the Khans' crowded home, however, and therefore rejects the Pakistani tradition that a bride joins the husband's household. He intends that both couples will live in his house, which boasts a carpeted bedroom and bath for each daughter. Enraged by George's hypocrisy in entertaining an offer that would separate two sons from the household, when he has declared his eldest son dead because he left, Ella confronts George and the visitor. With vigorous pride in her own working-class traditions of hard work and self-sufficiency, she defends her house and family against the implication that they are not good enough for this man's daughters.

When their visitor hastily departs amid insults from Ella, George turns on her and, exercising the only power left to him, begins hitting her. The threat of violence has hovered over preparations for the meeting: the children know their mother bore the brunt of George's anger when their older brother left. They, however, do not remain passive when the violence begins this time, but act in concert to restrain George and protect Ella. In the melee, Sajit suddenly vents his pent-up rage by madly striking George with his parka. Abdul proves stronger than George, and his assertion of strength redefines the family. Although he subdues his father, Abdul refuses to hit him. Afterwards he acknowledges that filial respect, as well as understanding of his mother's love for George, prevented him from striking. Nevertheless, Abdul places the family on a new footing grounded in British culture and says 'it's me dad that's gonna have to change' (74). When Abdul attempts to return Sajit's parka, which lies forgotten on the floor, Sajit shows that the changed situation allows him to dispense with this form of protection. He discards it in a rubbish bin.

East Is East demonstrates the incompatibilities between two cultural systems, as well as the factors linking them, through the dynamics within a family. The emergence of individualistic desires and goals that are considered a normal and necessary aspect of coming of age in Western societies threatens the primacy of family identity that George wants to impose on the family to maintain allegiance to Pakistani culture. The fact that he himself broke with such traditions to leave his country and marry an English woman seems only to have intensified his need to hold on to what remains of his cultural identification. George's concept of the family, however, is oriented towards the past, not towards the future of his children. Its assumption that the father's social power extends beyond the immediate family does not correspond to their reality. His attempt to pick out marriage partners for two of his sons exposes the limits of George's social power. Less affluent than the man with whom he is negotiating, George cannot insist that his sons and their wives remain within his household. This understanding of George's relatively powerless position gives the younger generation the ability to take control. As Abdul implies, they will reorganize the family along Western lines, and their father will have to accept the changes. The reorganization means they will not place family identity above individual identity. As Elaine Aston has observed, the play's action centres on the 'dis-identification' of the younger generation in relation to their father (*Feminist Views* 142). The play's resolution suggests that the family's engagement with Western individualism will preserve its closeness while limiting the oppressiveness of the father and allowing a greater range of choice for the children.

Tanika Gupta's *The Waiting Room*, first performed in the Cottesloe Theatre at the National in 2000, depicts the younger generation's achievement of freedom from the control of immigrant parents within a Hindu family. Priya's sudden death initiates the action. A middle-aged woman who emigrated from India and married the man of her choice, Priya has devoted her life to being a wife and mother. Though satisfied with the success of her professor husband and daughter, who is an environmental lawyer based in Paris, she feels guilt over her own lack of accomplishment and worries about her son's aimlessness and his alienation from her. The play uses a magical element in the form of a Hindu belief that the soul of a recently deceased person migrates through an intermediate spirit world before achieving ultimate liberation. Thus the deceased Priya appears among her family while they observe traditional rituals. A spiritual guide in the shape of her favourite Indian film star tells her that she must resolve conflicts in her earthly existence before she can be freed from it.

Priya reviews her life and finds significant conflicts. She reveals a secret she has long carried: a long-term affair with her husband's best friend, and his unknowing paternity of her youngest child, who died in a freak accident. Listening to conversations among family members, she also learns their secrets. Her daughter has a same-sex lover in Paris. Her husband and his best friend have resolved the tension caused by her infidelity. Her son Akash harbours deep, unresolved anger towards her. Focusing on repairing her relationship with Akash, Priya gradually understands that his anger stems from the profound sense of rejection he felt after the accidental death of his baby sister. Only six years old at the time, he lost control of the baby's pram while his mother shopped. Because of her own guilt and grief, Priya failed to comfort him and assuage his guilt, sending him away to relatives for a time. After many attempts to reach Akash in dreams, Priya finally succeeds in communicating her love. Akash can finally mourn her death, and Priya can release her hold on him and move into the realm of freedom beyond mortal life.

Priya's life encompasses elements of the typical immigrant experience as well as individual choices and events. She has not faced economic struggle, but experienced devastating loss. Priya transferred her own ambitions to her children, and those ambitions constituted a form of control that both resisted in different ways. Before taking final leave of life, she is able to give up the attempt to control them and liberate her children from the guilt that bound them to her expectations. By the end, Priya expresses her love as confidence in their future choices as fully formed individuals. The concept of the waiting room, though used in reference to the place that houses souls after death, also serves as a metaphor for a life devoted to nurturing the next generation. Living more for others than for herself, Priya has waited for her efforts to be rewarded. In the final reconciliation she ceases to wait and releases the next generation to form independent identities and make their own choices.

Blest Be the Tie by Doña Daley, first performed in the Theatre Upstairs as a joint production of the Royal Court and Talawa in 2004, focuses on the immigrant generation in an examination of identity. It presents a humorous and hopeful portrait of a Jamaican–English pensioner, Florence, who has lived in London since coming there as a young woman. Despite disappointments, Florence lives contentedly in her council flat and follows a weekly routine of activities with her best friend and neighbour Eunice, who is white. She takes pride in the success of her children and feels comfortable enough in her own British identity to criticize the Scottish dialect of someone she encountered on a train. When Florence's sister Martha, whom she has not seen since she left Jamaica 30 years ago, arrives for a surprise visit, she upsets Florence's comfortable identity and pride in her choices.

Each sister contradicts the other's expectations. Florence, who believes her family back in Jamaica remains as poor as they were when she left, finds Martha's evident affluence disconcerting and her expensive gifts embarrassing. Martha expects Florence to gladly accept her invitation to return 'home' to Jamaica, and finds her sister's refusal incomprehensible. Incompatibilities intensify, and the sisters' relationship almost breaks down completely when Martha discards Florence's worn but treasured sofa and replaces it with a stylish suite she has bought as a surprise. They argue bitterly about perceptions of and obligations to the family that changed after Florence left. They finally reconcile and vow to maintain contact, even though they now understand the deep differences between them.

This play overturns the common myth of disillusioned immigrants longing for a lost homeland. Like *The Waiting Room*, it demonstrates the more complex situation of a woman who did not find wealth or exciting opportunities in England, but who has made a life and found a sense of belonging in her adopted country. Florence realizes through the encounter with Martha that her true home is in England. Acknowledging that 'the road was more crooked than straight' and that she never quite knew 'what the prize was supposed to be' (66), Florence feels she made the right choice because her three children have achieved education, good jobs, and affluence. England has become her homeland not just because her children and grandchildren live there, but also because all her memories of her children's developmental years are linked to this neighbourhood, even though she was moved out of her much-loved house and into this flat. Florence also recognizes that the love she feels for her friend Eunice is like that of a sister. Despite their differences of race and place of birth, they have shared crucial life experiences, have understood each other deeply, and have integrated their friendship into the rhythms of their daily lives. The history she shares with Eunice contributes to Florence's sense of home. They have created a shared narrative and maintain it through habitual references to past events, as well as through the many photographs taken and displayed by Eunice. Now Martha's visit will become another episode in this ongoing story. Florence's achievement of identity and meaning cannot be separated from the place in which it has occurred.

With its all-female cast, *Blest Be the Tie* gives visibility and significance to the lives of the quiet and industrious British Afro-Caribbean women who have simultaneously created new homes for their families and transformed the communities in which they live, while remaining all but hidden in constructions of the nation. It emphasizes the specific, gender-based challenges faced by women immigrants. Subject to the traditional expectation that they will serve the extended family as well as their immediate one, they struggle with distance and financial constraints to attempt to meet the obligations inherent in this expectation. Serving the family usually means taking a back-breaking and low-paid cleaning job outside the home, but also encompasses responsibility for rearing children, keeping the house clean, and cooking meals. Florence exemplifies a woman who has successfully met these expectations and in the process facilitated the success of her children. She has never before looked back on her life. Martha's visit gives her a chance to do that, as well as offering the opportunity to go back to Jamaica and, in a sense, cancel the decision made years before. Florence's decision to stay in England is based not only on a sense of home, but also on a sense of accomplishment.

While the plays of Khan-Din, Gupta, and Daley exhibit a future orientation that forecasts assimilation into mainstream British culture and increasing ability to garner its rewards, a less optimistic tone characterizes Kwame Kwei-Armah's work. Kwei-Armah focuses on the British Afro-Caribbean experience by framing the contemporary Afro-Caribbean subculture within the African Diaspora and by emphasizing the ways in which it continues to be shadowed by the wrongs of slavery. *Elmina's*

Kitchen, produced at the National Theatre in 2003, came out of an effort by its new artistic director, Nicholas Hytner, to address the concept of nation implied in the theatre's name. Thus, the production of the play was a consciously political choice and part of a move towards heightened consciousness of the political in productions at the National. This political context highlights the position of *Elmina's Kitchen* in representing contemporary social conditions. The play takes a balanced viewpoint, celebrating black identity through family stories and music, contesting stereotypes of black men, and seeking public attention for an issue critical to the Afro-Caribbean community.

Kwei-Armah, a successful actor well known for his role as a paramedic in the BBC drama *Casualty*, was in his mid-thirties and the father of three when he made his playwriting debut with *Elmina's Kitchen*. Accordingly, his identification with Deli, the father who tries to keep his son Ashley from getting involved in crime, seems clear. Deli operates Elmina's Kitchen, a small West Indian restaurant in Hackney's 'Murder Mile', an area known for drug-related violence. The name of the restaurant carries a double resonance, as it not only honours Deli's mother but also recalls Elmina Castle, a fortress on the coast of Ghana that was used in the slave trade to hold captured Africans. The name thus creates a bridge between contemporary conditions for black families in urban England and the human injustice of the slave trade, and serves as a focus for the mixture of anger and pride in the legacies of the African Diaspora.

This realistic play involves three generations. A prominently displayed portrait of Elmina, along with a laminated poster proclaiming her philosophy of treasuring life, symbolically represents Deli's parentage. Clifton, the self-dramatizing father who long ago deserted the family, returns from Jamaica to reclaim relationship. Deli's son, 19-year-old Ashley, shows an adolescent's resistance to his father's authority and a need to assert his own manliness. Each generation faces struggle and danger. Clifton has lived for the moment throughout his life. Now old and suffering from Parkinson's disease, he pleads with Deli to take care of him and to go with him to Jamaica to 'show them my seed is something' (88). Clifton's earlier desertion stands between them, but even if it did not, Deli would have little capacity to help. All Deli's energy goes towards suppressing his own rage and maintaining a stable environment for Ashley. Even the sexy and caring Anastasia, who takes a job in the restaurant and offers Deli love, does not distract him from his vigilant control. Having survived a stint in prison, Deli avoids drugs and crime, and he is determined to keep Ashley away from them too. His brother's murder in prison just before his scheduled release intensifies Deli's resolve to protect Ashley from such a fate.

Ashley, however, seeks autonomy and sees power in the guns, cash, and expensive cars of the neighbourhood gangsters.

Deli's hard work, optimism, and discipline ultimately fail to prevent Ashley from getting involved in a gang. When he learns that Ashley participated in the beating and arson aimed at intimidating the owners of a rival restaurant, he informs the police, feeling that this desperate measure offers the only hope of separating Ashley from the gang. Deli negotiates with the police to let Ashley off in return for testimony against the gangster Digger. Before he can get Ashley to a safe house, however, Digger shows up. Preying on Ashley's shame that his father turned informer, Digger gives Ashley the job of killing Deli. Ashley seems about to do so, but Digger unexpectedly kills him, leaving Deli shattered by the loss of his son.

This play's view of gangsters and their ruthless violence, in contrast to that of the in-yer-face plays, gives tangible form to history, culture, and identity. Evident in the name of the restaurant, the photograph of Deli's mother, the food, the music, and the stories, shared history and culture nurtures and strengthens. Outside the oasis of the restaurant, however, their shared identity presents obstacles to the survival of all the characters. The danger is most acute for the young man burdened with the deferred dreams of two generations. The play's sympathetic characters and its intimate view of their struggles argue against the detached amorality of the in-yer-face play. *Elmina's Kitchen* employs a moral framework and appeals to a social conscience. Ashley dies in full view of the audience, having, in his fit of youthful rebellion, turned against the father who loves him enough to sacrifice everything for his safety. Ashley's death touches the audience with its tragedy.

Elmina's Kitchen resists a sense of futility in the death of this young man, implicitly appealing for the prevention of such tragedy. Ashley's family life contained important factors that could have protected him from crime and early death. His father had overcome a lack of education, a prison record, and the break-up of his marriage; he parents his son responsibly and serves as a stable and law-abiding role model. Deli supports Ashley's schooling. Social factors, however, influence Ashley to denigrate his father's hard work and stability because they do not offer the street status and quick money that seem to constitute freedom and power. Until he recognizes Ashley's vulnerability to the appeal of gangs, Deli has avoided conflict with the local gangsters by allowing Digger to hang out at the restaurant and claim him as a friend. Ashley's fascination with Digger, combined with the shock of his brother's death and the hope that the restaurant makeover initiated by Anastasia might improve their situation, leads to Deli's decision to cut Digger out of his life. That decision, rather than protecting Ashley, leads directly to his death, as Digger, who had previously refused to admit Ashley into the gang, retaliates by taking him in and involving him in a crime sure to throw suspicion on his father. Clearly, the power of criminals in this community has made tragedy, if not inevitable, very common. The fact that Anastasia has also lost a 19-year-old son reinforces the sense that, in Clifton's words, 'all you generation curse' (89).

While it does not provide an explicit answer to the question of what has cursed young black men in Hackney, the play does point to forces beyond the control of the family at the centre of the drama. Ashley, unlike the young women of Prichard's *Yard Gals*, does not lack a home and caring parents. He gets caught up in crime because it is ubiquitous where he lives, and because the Yardies project an image of West Indian–identified masculine power that lures young men. *Elmina's Kitchen* takes its audiences into the heart of an actual family in the midst of the urban crisis created by gangs. In the face of appalling statistics regarding lack of education, unemployment, and criminal convictions among young black men, it argues against assumptions that prompt dismissal of such statistics through the human specificity of a father and son. Premiering at the National Theatre and going on to play in the West End, this play challenged public indifference towards crimes in which both perpetrator and victim are black in 'Murder Mile' and other poverty-afflicted areas.

Kwei-Armah followed *Elmina's Kitchen* with *Fix Up* at the National Theatre in 2004. Like *Elmina's Kitchen*, *Fix Up* is written in realistic style and takes a complex view of individuals involved with the ongoing creation of Afro-Caribbean culture in Britain. This play also focuses on a small business, the Fix Up Bookstore operated by Brother Kiyi in Tottenham in north London. Devoted to books about black history, philosophy, and culture, the Fix Up Bookstore represents a dream of community more than a dream of economic progress. Brother Kiyi cares little about turning a profit, as he lends books without charge and spends what little he takes in on acquiring rare books. Though he has few customers, he occupies a respected place in the community because of his knowledge. Kiyi is single and without children, but the shop's regulars constitute a kind of surrogate family. They include Carl, a young man whom Kiyi is teaching to read; Norma, a barber and part-time preacher who plays draughts with Kiyi and uses her savings to try to rescue the shop; and Kwesi, an activist who stores mysterious chemicals in a room above the shop.

Development of the area threatens the bookstore, because the building in which it is located is slated for remodelling into luxury flats. Plans reportedly call for replacing the bookstore with a shop selling hair products. Kiyi refuses to even acknowledge the threat, but the arrival of a new customer, Alice, whose beauty captivates the male regulars and whose appearance announces her mixed racial heritage, disturbs his enclave. In a series of visits, Alice converses with Kiyi, buys a few books, and becomes fascinated by a volume of slave narratives titled 'Family' in the bookstore's stock. She also notices a locked box and opens it when Kiyi is absent. The photo album inside confirms what Alice has evidently suspected – that Kiyi is her biological father, who gave her up for adoption when she was a toddler. In the confrontation that ensues, Kiyi's constructed history crumbles. He admits that the 'Family' narratives are fictions written by him and that Alice was taken from him because he killed her mother. Carl and Norma witness Kiyi's unmasking. In its aftermath Carl reveals that the boxes Kwesi has been storing in the room upstairs contain hair products for the shop Kwesi plans to open when the bookstore closes.

Fix Up addresses both the strengths and the weaknesses of the black consciousness movement that developed in the 1970s. Clearly, the movement enabled Kiyi to find meaning in life after the devastating loss of his family through the accidental killing of his wife and removal of his child. Through his teaching and example, Kiyi has encouraged Carl to learn to read and inspired Norma to become active in local politics. Kiyi has, however, failed to communicate the profound source of meaning he finds in black history to most of the people surrounding him in his predominantly black community. Despite the fact that it is Black History Month, few people come into the shop and almost no one buys a book about black history. In valuing abstract ideas, Kiyi barely acknowledges physical necessities, and would have no food if not for the donations brought by the more practical-minded Carl. The people around him, by contrast, typically concern themselves with immediate needs rather than history. They exercise identity through desire for concrete symbols of status, rather than through intellectual understanding of power. That is why the shop will soon be replaced with luxury flats and a shop selling hair products. Even Norma demonstrates the need for visible signs of status. Though she is a willing student of Kiyi's philosophy, the first thing she does in preparation for entering a political contest is to buy a new wig.

A serious failure of the Black Power movement becomes clear in the interaction between Kiyi and Alice. It romanticized a notion of full-blooded blackness that denied the reality of mixed heritage in the African Diaspora. Alice feels that living with her white adoptive family has deprived her of a true identity. She comes to the bookstore to learn about the unknown aspect of herself, but Kiyi's version of black history does not allow her to integrate the separate parts of her heritage or find her lost family. The abstract ideas voiced by prominent figures of the past on Kiyi's constantly playing tapes have no meaning for a younger generation that has been deprived of familial nurture. Kiyi has substituted the heroic histories and rhetoric of black leaders for the personal history he denied his daughter. He has sublimated his own longing for family in a fictional history of families. His dishonesty angers Alice, who says, 'what did I find, a sad, old, hateful man who pretends to unearth the truth, but houses lies' (73).

Resignation replaces bitterness as Kiyi and his friends box up the shop's contents and Alice visits for the last time. Each character now understands the necessity of self-reliance, without the support of a community but with some token of what that community might have represented if it had ever actually existed. As Norma and Carl pack books, they recite passages from writing by Maya Angelou and James Baldwin, indicating that they value what they have learned and have made the thoughts of these black writers part of their own thinking. Kiyi cuts off his dreadlocks in a gesture of loss, and intones a chant that evokes the sorrow of slavery. He gives Alice the album, offering reconciliation, and she acknowledges that the painful excavation of the truth has given her the foundation for a more authentic identity. As the play ends, Kiyi's 'castle', as Alice refers to his construct of a black community, has fallen. In its place is an empty space that signals both absence and possibility. This emptiness or openness attests to both a potential for collective power and an uncertainty about what form it might take. The sense of possibility, while shadowed by loss, reverberates in the empty shop. However, the knowledge that the space will soon be occupied by a niche business targeting black people serves to temper idealism.

The production of work by Afro-Caribbean and Asian dramatists and the increasing visibility of non-white actors in mainstream theatre provide evidence of the multicultural nature of British society. Plays by Williams, Khan-Din, Gupta, Daley, and Kwei-Armah explore well-documented changes not only in the lives of immigrants and their children, but also in the nation that has become their permanent home. These plays redefine national identity as much as they articulate issues of individual identity. They make visible what has occurred but has not been fully recognized: the development of a heterogeneous cultural landscape in which there are various patterns of connection and separation. These plays exemplify contemporary multiracial, multiethnic Britain and its continual renegotiations of separateness and integration. The plays that combine realism with magical elements evoke the intercultural nature of post-Thatcher Britain. The cultural consciousness of these plays is also evident in other types of recent theatre, such as the National Theatre's 2003 modern-dress production of *Henry V*, which featured Adrian Lester in the title role, and the 2005 ska musical, *The Big Life*, produced by the Theatre Royal Stratford East, that celebrated the arrival of West Indian immigrants on the Empire Windrush in 1948.

Generational plays by dramatists of Afro-Caribbean and Asian backgrounds interpret recent developments in British society against a background of immigration, aspirations for better lives, and generational difference as the immigrants struggle to adapt while their children grow up with a set of assumptions and expectations derived from the new environment. All the plays represent the immigrant generation as hard-working and productive, but only Daley's *Blest Be the Tie* shows an immigrant able to participate fully in English socio-cultural life. Young people in this set of plays question the values of their parents and often throw off the constraints of their control, but encounter various difficulties in pursuing their own directions. The most serious social issue the plays make visible is the detrimental influence of a criminal culture that profits by exploiting the young as drug users and sellers. Young people in families with poor economic prospects and uncertain social position may feel a heightened need for a sense of belonging, power, and status, and thus prove particularly vulnerable to the appeal of the gangster image.

Though their plays make visible the presence of the communities in which they are set and give a central place to their particular political issues, Asian and Afro-Caribbean playwrights make it clear that multiple factors shape the lives of individuals. Ethnic and racial identification interacts with such elements as location, economic position, and education, especially as these factors change between one generation and the next. The ability of characters of one race to adopt the culture created by another, as is the case with Alice in *Fix Up*, Tone in *Lift Off*, or Eunice in *Blest Be the Tie*, negates any essential connection between race and culture. Characters such as George in *East Is East* or Kiyi in *Fix Up*, who define themselves and others in terms of an inflexible code, ultimately fail to compel allegiance to this code or persuade others of its validity. Characters such as Abi in *The No Boys Cricket Club* or Priya in *The Waiting Room* who find a way to situate individual identity in both personal and collective narratives do so through reviewing and questioning their history. They acknowledge both past and present in their choices and ultimately understand the way in which their identities are anchored in history. The plays written by the descendants of immigrants, in contrast to those of the first in-yer-face playwrights, communicate an interest in the future and a sense that they stand at the beginning of a history yet to be written rather than at the end of one that has been discredited.

Generational transition and the post-Thatcher working class

Class divisions, which have created and reinforced patterns of inequality since the beginnings of the modern nation, present new issues in the post-Thatcher era of industrial failure, multinational corporations, unemployment, and outsourcing. Plays about working-class life thus constitute an important current in the resurgence of political drama. They arise from a rich background of dramatic writing from the late 1950s to the present (see Lacey, 1995) and draw on the theatres of working-class enclaves in the north of England and in Scotland. Companies such as Hull Truck, founded in 1971 with the mission of providing excellent theatre accessible to local audiences, and Newcastle's Live Theatre, started in 1973 to develop and attract new audiences in the Tyneside area, have produced a steady stream of plays addressing working-class communities. From such companies came work by C.P. Taylor, John Godber, Willy Russell, Tom Hadaway, and Phil Woods. While the explicitly political touring companies such as 7:84 and Northwest Spanner, which helped to define the political theatre of the 1970s and 1980s, have not survived, the theatres that have established a loyal local following continue to create new work. Moreover, the increased linkage between regional theatres and London theatres, and among regional theatres, has made the work of playwrights who develop a regional voice accessible to a geographically widespread audience.

Plays that focus on the working class in contemporary society make visible the conditions faced by communities and families associated with Britain's crafts and industries. The generational dialogue has proven to be a particularly fitting context for presenting working-class concerns. The working class bore the brunt of Thatcherism's restructuring of the British economy and has been most harmed by changes in British society since 1980. New plays that reclaim the place of working-class people in British political consciousness revolve around one incontrovertible fact: the working class still exists, but their customary forms of work are disappearing. Loss of work implies the loss of not only economic power but also personal identity. It often brings about the destruction of communities and distinctive ways of life. This loss and destruction contributes to the major break in continuity between the pre-Thatcher and the post-Thatcher generations of the working class.

The economic and political changes that have altered the landscape of working-class Britain in the latter half of the twentieth century constitute a major element in the acknowledged history of the era. Thus plays focusing on the working class, like those that focus on the immigrant experience, maintain a strong connection with history and often embed this sense of history in a generational dialogue. They attempt to make known people and occupations that, even when they were central to Britain's economy, were relegated to the margins of cultural representation. The transitions at the heart of these plays centre on a family or closely connected group of individuals. The plays dramatize change through the classic political styles of social realism and Brechtian epic theatre. They use the medium of drama to, as Peacock has suggested of the 1970s dramatists, 'communicate the private experience of public events' (1991: 79), or show the personal impact of social and economic changes.

In *Skinned*, co-produced in 1998 by the Nuffield Theatre Southampton and the Chelsea Theatre Centre, Abi Morgan presents a disturbing picture of community division and disintegration. The play's setting in an ornately tiled Victorian-era slaughterhouse that is slated for demolition signals the abandonment of a traditional occupation. Adjoining this hall offstage is a modern meat-packing plant with a mechanized production line. The plant provides employment but requires little strength and skill from its workers and gives them no distinctive identity. The plant workers make their initial appearance in the bloodied white uniforms of the job. Joe and Eileen, the oldest members of the group, had worked in the old hall where animals were slaughtered in the traditional way. Joe is a master butcher, and Eileen a meat inspector. Rose, who is in his mid-twenties, works as a stunner, shooting each animal with a stun gun before it is killed. Seventeen-year-old Shelley processes meat on the line, and sixteen-year-old Mat has just begun to work at the plant.

The workers use the empty hall to take a break from their jobs and pursue personal interests. Eileen helps Shelley prepare for her wedding with Rose, to be held in the old slaughterhouse. Joe helps care for Eileen's son and takes every chance to be alone with her. Rose, who leads a group of local lads known for fighting and hooliganism, plans the evening's activities and invites Mat to join them. Racial tension emerges as the white workers denigrate an absent black co-worker and express irritation over music audible from a nearby black club. Their conversation gradually reveals that the black co-worker's teenaged son, who used the hall for gymnastics practice, was beaten to death there recently. The murder has not been solved, but the police suspect Rose. The wedding, occurring so soon after the murder and in the same location, crystallizes the racial conflict. On the eve of the wedding, Rose's former girlfriend Tam, who has left the community and become a secretary in London, returns. She doggedly pursues an agenda that remains unclear until the play's climactic scene reveals that she has returned to reclaim Rose. Rising racial tension culminates in an offstage race riot. Onstage, Shelley gets drunk and, prodded by Tam, confesses to the murder, exonerating Rose and permitting Tam to claim him.

This play deconstructs working-class masculinity and reverses gender roles. *Skinned* strips skinheads of their ferocious image and exposes them as weak, frightened, and dependent. Despite his knife-brandishing bravado, Rose cannot even maintain an aggressive pose. In the face of threats from the nearby club, he breaks down, surrenders his knife to Shelley, and begs her not to tell the lads about this failure. Late in the play, he confesses to the police, in an attempt to save Shelley, but neither she nor Tam expects him to sustain this action. Moral strength has disappeared along with masculine pride, as is revealed when Joe admits he was present at the murder but did nothing to stop it. By contrast, the women demonstrate their hardness. Tam pursues Rose relentlessly, using the bereaved black father as a pawn and pausing only to consider whether publicity over the murder might threaten her job. Shelley admits that, though she joined in beating the black teenager only in response to a teasing suggestion that she fancied him, she enjoyed a thrill when she landed the most vicious blows to his face with her boot. The men's disempowerment and impotence seem to have generated aggression and violence in the women.

The pivotal character Eileen frames the play's viewpoint. Her monologues introduce and conclude the action. Present, though often unnoticed throughout, she records and reacts to the assertions and admission of those around her, serving as a substitute mother to Shelley while someone else minds her offstage child. Eileen's strategy of not asking questions has helped her to avoid knowing too much, but when she learns that Joe witnessed the murder and did not intervene, she expresses shock and condemnation. Her adaptability surfaces in concern about practical things, like the unused food for the wedding reception. Alone at the end, she tries to come to terms with the murder of one young person, incarceration of others, and the destructive race riot. Eileen's series of wordless attempts at situating herself communicate uncertainty about how to interpret or respond to the events. Nevertheless, her position as framer of the narrative suggests that she has begun not only to adapt to circumstances but perhaps to confront and attempt to improve them.

Trust by Gary Mitchell, produced by the Royal Court in association with the National Studio in 1999, uses the failure of working-class life as the context for a suspenseful conflict in a Loyalist family in Belfast, Northern Ireland. Geordie and his friend Arttie are active in the local Protestant paramilitary, but except for one scene in which they bargain with someone selling stolen guns, they are idle, watching television at home or drinking at the club. When Trevor, an acquaintance just released from prison, comes to ask for help in getting work, Geordie responds sarcastically, but adds earnestly, 'If I could get a job for anybody I would get it for myself' (17). Geordie's wife Margaret, in contrast to the men, juggles different roles and forms of work. She supports her family and provides food to neighbours in need, while serving as her husband's confederate in his paramilitary activities and keeping an eye on their teenage son Jake.

Trust, as the play's title indicates, forms the primary concern of this family. Its public aspect appears in the rough and thorough check for listening devices given to Trevor before Geordie will speak with him about paramilitary activity. The private aspect emerges in the conflict between Margaret and Geordie over Jake's avoidance of school and Margaret's conviction that he is being bullied. She asks Geordie to intervene on Jake's behalf, but he insists that Jake learn to ward off bullies on his own. The timid Jake does not show himself an apt pupil, but Geordie is preoccupied with the deal to buy guns. Meanwhile, Margaret takes matters into her own hands and arranges with Trevor to rough up the bullies. Trevor's involvement escalates the situation, and Jake ends up under arrest after stabbing one of his tormentors. Jake's arrest forces Margaret into making a choice between husband and son. When she chooses the son, she also betrays her family's Loyalist politics. Desperate to save Jake from prison, Margaret turns a gun on Geordie, trying to force him to bargain with the police to get the boy off. When Geordie leaves, Margaret goes even further, phoning the police with an offer to inform on her husband in return for Jake's freedom.

The family's confinement in a situation that constrains choice eventually destroys their essential bonds of trust. Lack of employment has left Geordie with only the dangerous and illegal paramilitary activities as a source of status and sense of purpose. As if they were workmates, Geordie spends far more time with his friend Arttie than with his wife or son. Dependent on the hard-working Margaret, Geordie nevertheless rebels against performing traditional wifely tasks like making tea. By insisting that Jake learn to fight, he opposes his wife and attempts to recapture a traditional fatherly role in relation to his son. While Geordie invests his energy in fantasy scenarios of weaponry and strength, Margaret copes with daily reality and uses violence only when necessary. Unromantically, she concludes, 'Husbands can be replaced. Sons can't' (42). The young Jake faces the most painful dilemma, as he moves towards maturity and thus towards inevitable and irresolvable conflict. The manhood awaiting him, with its football, beer, and violence, frightens and repels him. Jake tries to hide, but the situation affords no escape for anyone. Seeing his mother betray his father exposes him to his own powerlessness. As Russell observes in regard to other works by Mitchell, the play moves 'from communal solidarity to personal loneliness' (197). Mitchell sidesteps the usual themes of plays about Northern Ireland to highlight joblessness and economic deprivation within one working-class community.

Michael Wynne's *The People Are Friendly*, first performed in 2002 at the Royal Court Theatre Upstairs, observes a community at the point of transition, when a way of life that has existed for generations is erased. Set in Birkenhead, a city on the northwest coast of England long dominated by shipbuilding and shipping, the play shows the effects of the shipyard closings and redevelopment on four generations of one family. The play begins with deceptive simplicity and the appearance of a success story, as Michelle plans a barbecue to which she has invited her parents, her sister Donna, and Donna's husband, 7-year-old son, 16-year-old daughter, and infant granddaughter. After several years in London, Michelle has returned to Birkenhead to take a new job and renovate a large Victorian house near the estate where she grew up. Michelle's plans begin to go comically awry when her partner Robert proves inept at firing up the charcoal grill. As her family arrives, Michelle loses control of the situation she had hoped would be a triumph.

Comic collisions involving cheap wine and stolen tequila, bluntly rejected canapés, breast surgery, drug dealing, a dead cat, fears of tainted meat, a fight at the local pub, and Robert's departure engulf the gathering. Against this backdrop, Michelle finds herself seriously at odds with her family. The distance between mother and daughter becomes poignantly clear when Michelle's mother sets to work cleaning a spot from a rug as if she were a hired cleaner. She actually misses her cleaning work, which she quit out of deference to her husband's pride when he lost his job. Michelle's father arrives at the party with the surprise announcement that he walked out of his current job as a parking attendant because activity at their old workplace has convinced him and others that it will reopen and they will be rehired. Michelle, who has concealed the fact that her new employer is the company that formerly employed her father, must make the agonized admission that she has been hired to help manage the redevelopment of the site. The shipyards will not reopen, but will be replaced by upscale housing, shops, and tourist attractions.

A passionate argument between Michelle and Donna ensues. Like the sisters in Churchill's *Top Girls*, they represent conflicting interests in a Britain divided between haves and have-nots. Michelle tries to present the redevelopment as a positive effort to preserve the site's history and offer new kinds of jobs to the locals. Observing that this type of development benefits only the rich, Donna bitterly suggests building a marina for yachts with a highway bypass to keep the less affluent out of sight. She predicts that the new jobs will be temporary, part-time, and unskilled. Michelle urges her sister to be realistic. In turn, Donna describes her own reality of trying to keep her family from sliding farther into dysfunction while 'working nights and split shifts packing fuckin' peanuts', holding off financial catastrophe by 'paying one loan off with another' and borrowing small sums from her mother to make it to the next pay day (119). Though she wants to blame Donna for a lack of energy or ambition that kept her in Birkenhead, Michelle instead admits that she envies Donna's identity and sense of purpose. As their distraught parents leave without touching the food, Donna faces a moment of decision in whether to go or stay. She too departs, signalling to Michelle the impossibility of maintaining family closeness as long as she is employed in destroying the last vestige of hope for reopening the shipyards. Michelle has urged adaptation to new economic circumstances but now, as Donna's curt farewell tells her, she will have to adapt to a changed family. Alone and crying in her house, Michelle hears a small noise that alerts her to the fact that she is not quite alone: the infant has been left behind. Unexpectedly, she cuddles and comforts the baby.

Dramatizing the transition from an industry-based to a service-based economy, *The People Are Friendly* shows Michelle succeeding financially while the other members of the family are falling farther behind. Although it does call attention to the economic aspect of recent changes, it focuses even more on the less tangible loss of pride in doing something worthwhile. This sense of pride has linked Michelle and her father, but now it divides them. John's pride arises from his shipbuilding skills, while Michelle's is based in the essentially social skills, like being friendly, that lead to success in the service economy. As representatives of their respective generations, the father and daughter pursue similar goals of personal independence and stability, and both try to contribute to their family and community through their work. Economic changes, however, have opened a path for one that closes off possibilities to the other. Though objectively Michelle cannot be faulted for trying to make the best of the opportunities available to her for advancement, the family's sense that their losses enable Michelle's success cannot be dismissed. While she had hoped, however unrealistically, to renew family ties, her participation in the company that has made the shipbuilders redundant defines her as one of the enemy.

The break with her family has major significance for Michelle because, as she has conceded in her argument with Donna, success means little to her outside the context of her family. She has wanted to make them proud of her. Instead, she has brought into their midst the social conflicts engendered by economic transition. As she vents her own sense of loss, the unexpected presence of the baby negates closure of the action. The playwright specifies that the infant left behind in the hasty and emotional departure is 'a real tiny baby' (126). The baby – not just realistic but inescapably real – represents new beginnings, the future, and hope. Michelle holds this real person and evocative symbol in her arms and says, 'I'll look after you now' (*ibid.*). The question of how Michelle and the other post-Thatcher managers of the new economy will 'look after' the working class of the future constitutes the question that hangs over this final moment.

Richard Bean's *Under the Whaleback* (2003), first performed at the Royal Court Theatre Upstairs, traces the decline of the fishing industry and its way of life through three generations of men. Structured in three scenes dated 1965, 1972, and 2002, the play's stage directions suggest role doubling to connect the changing ages and relationships of the characters. Action occurs on three different fishing vessels represented on stage by a bunk-lined crew's cabin in the forecastle beneath the whaleback – the upward-curving deck at the bow – of 'a fifties built eight hundred ton sidewinder trawler' (18). The trawlermen speak in the specialized language of their work and the accents of their home city of Hull on England's northeast coast. The shared frame of reference invoked in their conversations includes pubs frequented by the seamen and housing estates in the vicinity. Darrel, born in 1948, moves through all the scenes. In the first, as a 17-year-old 'snacker' or apprentice deckhand making his first run on the Kingston Jet, he meets an old hand named Cassidy, who instructs him in the terminology and procedures of the job. Cassidy surprises Darrel by claiming to be his biological father, pointing out some shared genetic traits to bolster the claim. Anticipating his own death, Cassidy gives Darrel an inheritance: a waterproof survival suit and instructions on where to find a hidden betting ticket worth £300. In the second scene, Darrel has become a seasoned trawlerman, and Cassidy, seven years dead, has become a legend among the seamen. Darrel and other members of the crew – Bill, Roc, and Norman – wait out a fierce storm below deck on the James Joyce. The men discuss Cassidy's death, which occurred when he left a life raft and swam back into a sinking vessel seemingly on a whim. The storm eventually turns the trawler over on its side just after Darrel has donned storm gear and gone above deck for food to silence Norman's childish complaints of hunger.

The final and longest scene takes place in the present in a dry-docked trawler that has been converted into a museum. Darrel, who is now middle-aged and moves slowly because of an old injury, works there as a curator. As he closes up for the day, Pat, a younger man, comes in and begins to ask him about his shipmates, identifying himself as the son of Roc, who died in the 1972 storm. He has just been told of his parentage by an uncle. A cigarette smuggler who alternately smokes and takes drags of an asthma inhaler, Pat begins by asking friendly questions and offering Darrel cigarettes. Darrel tells him that his last run on a trawler was in 1975, and that he has been unemployed until the recent opening of the museum.

Pat soon gets to the point of his visit: his uncle has told him that Roc was one of the two men who got into a lifeboat, along with Darrel, in the storm that sank the ship. Pat wants to know how Darrel survived while the other two died. Darrel explains that he was dressed in his waterproof suit because he had been on deck, while the others had been below and not dressed to withstand water and cold temperatures. Angrily dissatisfied with this explanation, Pat uses the ruse of a card trick to get Darrel to spread his hands out flat, then nails his hands to the table with a pair of nail guns brought along for the purpose. Menacing the immobilized Darrel with the nail guns, Pat demands to know more. Darrel finally directs him to a picture of the entire crew of the James Joyce in an adjoining room. Looking at it, Pat realizes that Norman, rather than Roc, was his father.

Darrel reminds Pat of the many men who died at sea, commemorated on a plaque in the museum, and then relates the full story of Cassidy's death to illustrate the difficult life of seamen. He describes Cassidy's death as suicide and says that the betting ticket given to him by Cassidy had a map on the back that led him to a site where twin newborn sons had been buried wrapped in a trawlerman's oilcloth coat. Darrel speculates that Cassidy killed his sons, while letting his three daughters live, because he wanted to end the string of deaths at sea, that had claimed every male since his great-grandfather. Pat calls an ambulance, but before it comes Darrel's nine-year-old daughter arrives to fetch him home. Since Pat quickly conceals Darrel's nailed hands, Elly chatters with naive enthusiasm about her grandfather Cassidy's statue in the museum and her ambition to 'be a trawler skipper' (99) when she grows up.

In *Under the Whaleback*, as in his 1999 play *Toast*, Richard Bean provides a vivid window on the work that goes into producing commonly available consumables taken for granted by the public. The play focuses on the discomforts and dangers of this traditional occupation. As Cassidy tells the young Darrel, working on the trawlers is 'the worst fucking job in the world and only those what is born to it, what is gorrit in the blood, can do it' (22). While he emphasizes the toughness and occasional heroism demanded by life on a fragile vessel in arctic waters, Cassidy acknowledges that their extreme effort produces nothing more significant than 'the fish half of a fish and chip supper' (28). Nevertheless, he derides the more comfortable jobs available in the city, such as building caravans and fitting carpets. Cassidy relishes his own wild reputation, based on episodes of riding horses into pubs, launching fireworks from his posterior, and punching the Archbishop of York. Subsequent developments, however, reveal the depth of ambivalence Cassidy feels regarding the occupation that has defined and ended the life of men in his family for generations.

Darrel spends only ten years working on trawlers. Presumably the injury that affects his movement in the third scene ended his days at sea as the fishing industry went into decline. The discovery of Cassidy's dead sons, coupled with his experience of barely surviving three days in a lifeboat after the storm in 1972, has reconciled him to the loss of his occupation. Rather than having a store of colourful tales to tell, Darrel offers the simple statement, 'I'm still alive' (74). Though quietly bitter about his 25 years of unemployment amid the increasingly evident outsourcing of jobs, he seems to have adapted to the fact that his previous way of life has become history. Darrel has gained some education, married a teacher, and moved away from Hessle Road, the enclave of families involved in the fishing industry. Conscious of his heritage, he now finds purpose in educating mostly young people about deep sea trawling and the men who spent their lives in this line of work.

The negative effects of the decline of a once powerful industry appear in the resentment, frustration, and ennui of the younger character, Pat.

Though Pat seems energized by the idea of discovering his father, and enjoys stories about Cassidy, he scorns the idea of educating himself about the history of his father's occupation. Nevertheless, what Pat refers to as his 'vocation' (74) of smuggling cigarettes unconsciously mimics aspects of the trawlermen's life: frequent trips across the sea, but on a ferry to Belgium, and the constant risk of arrest rather than drowning. Just as a previous way of life shaped the men of Hull, Pat is being shaped by the way he earns money. He easily strikes up an acquaintance, quickly broaches the question of whether Darrel received a financial settlement after the shipwreck, and methodically employs violence to get what he wants. His restless dissatisfaction keeps surfacing, however, and eventually Pat reveals his envy of the men whose world offered them little choice and demanded little decision-making. Like the characters of Ravenhill's *Shopping and Fucking*, he considers freedom a burden. With his extravagant self-pity, Pat shows little personal strength, physically or emotionally, but exhibits a kind of toughness in his lack of empathy for the pain he inflicts. Oblivious to Darrel's nailed hands, he asserts at one point, 'I've treated you with respect haven't I' (87). In a brash bid for recognition, Pat concludes the bizarre encounter by placing a business card in Darrel's pocket.

Although it deals realistically with the material conditions of Hull working men, the play also conveys a sense of the almost supernatural attraction that the life at sea exerted on previous generations. Those who adapted to the life – Cassidy, Darrel, Darrel's crewmate Bill, and even the novice Roc – seemed to draw an extraordinary strength from a kind of union with the sea. Spending most of their lives at sea, with only short stretches on land where they father children they may never get to know, the seamen evoke myths of the selkie folk – enchanted seals that could take human form and mate with humans. These mythical beings were never happy with life on land, no matter how pleasant, and always contrived to return to the sea. They sometimes tried to claim their half-human children, but this inevitably brought tragedy. The forms of longing evident in the jaded, nihilistic Pat and the innocent, eager Elly cannot be separated from their heritage as children of the trawlermen. Both are children of division – not only the division between traditional ways and contemporary life, but also the division between land and sea. Exemplifying the future of their families and community, the two compel thoughts about the particular difficulties and risks they will face and whether they will survive them.

Lee Hall, who has worked for decades in Newcastle's Live Theatre, has taken northern working-class life as a consistent focus. Influenced by

John Godber and Willy Russell, with their humorous and popular portrayals of working-class characters and themes, Hall's most important
work to date is *Billy Elliot* (2005). Originally created as a screenplay
produced by Universal Pictures and released in 2000, and more recently
adapted into a large-scale musical – a form not often associated with
political theatre – *Billy Elliot* addresses changing conditions within the
British working class while it salutes the enduring spirit of the miners. Set
in 1984, at the beginning of the miners' strike, the action encompasses
the desperate stand of the miners who wanted to preserve their jobs and
their dramatic defeat by a Conservative government determined to
crush the unions. The epic-style dramatization and the panoramic stage
give the miners' community a presence in the musical that is not realized
in the film. The town hall and other public spaces, along with a large
ensemble of actors, represent the Geordie pit village. The attacks and
counterattacks that, to Hall when he was growing up in a mining town
during the 1980s, seemed 'akin to a civil war' (Billy Elliot programme,
2005: 2) appear in a series of vivid and large-scale images such as a
Christmas party with a giant panto-type caricature of Margaret
Thatcher and a line of police in riot gear. A version of the traditional
labour anthem 'Solidarity' expresses the unity of the miners, and the
hymn-like melody and lyrics of 'The Stars Look Down' invokes the
heightened emotion of the conflict. Dance sequences in which the entire
ensemble of about 40 adults and children form definable groups, weave
in and out among one another, and dissolve into different groups,
provide a visual representation of the conflicts, shifting loyalties, and
negotiations taking place in the community.

The miners' fight to maintain their livelihood frames Billy's personal
story. Billy attends Saturday boxing classes but has no aptitude for the
sport. Instead, he finds himself drawn to the ballet class that follows
boxing. Soon he pays his weekly 50 p to the dance teacher, Mrs.
Wilkinson, rather than to the boxing coach. Billy overcomes initial self-
consciousness, as the only boy, and shows extraordinary talent, with the
music underscoring his sense of discovery and growing pleasure in ballet.
Mrs. Wilkinson, despite a wariness of emotional involvement with her
pupils, provides a partial substitute for Billy's recently deceased mother.
She encourages Billy to audition for the Royal Academy of Ballet. Billy
wants to audition, but does not know how to express such a desire to his
family, where every male is expected to become a miner. When his father
discovers that Billy has dropped boxing in favour of dancing, he forbids
Billy to continue ballet. Billy's frustration, anger, and grief pour out in a
solo dance sequence that begins with jerky and percussive movements

evoking impotent rage and moves to a confrontation with a phalanx of policemen holding riot shields. Billy leaps against and batters the shields in a single-handed and futile battle to remove the obstacles to his family's well-being and his own desire.

Being cut off from ballet deprives Billy of his primary joy, but he finds two important sources of comfort. A letter left to him by his mother assures him of her continuing love and encourages him to be himself. His friend Michael urges Billy to express his individuality, showing his trust in Billy by revealing the homosexuality and enjoyment of cross-dressing that set him apart. When Christmas arrives in the midst of the difficult times, the miners hold a party, rallying their spirits in a satire about Maggie Thatcher as 'privatizing Santa' (Programme 18). Unable to cheer his son, Billy's father begins to comprehend that dance is vitally important to the boy, but he simply does not understand why. Billy explains in the song 'Electricity', evoking freedom and transcendence as he dances with an intensity that culminates in being flown above the stage. Billy's father agrees to let him audition, but must now find money for the trip to London. The strikers show their solidarity by taking up a collection, but only one of the despised scabs has enough money to contribute the needed funds.

Accompanied by his nervous father, Billy braves the alien world of the audition and returns to await the decision. The family and community wait along with Billy, but the letter's arrival is upstaged by the collapse of the strike. Alone when he reads of his acceptance to the Royal Academy, Billy at first hides the letter and tells his family that he has failed to get in; however, when he sees their disappointment he understands that they genuinely desire this opportunity for him. His family's support enables Billy to leave them. He stands clutching his suitcase as the men of his family and town, in full miners' gear, enter an elevator and disappear beneath the earth, singing of pride, unity, and work. In the final moment, Billy sings a farewell to his dead mother.

This musical, which, since its opening, has won awards and brought in full houses, presents its unabashed protest against the Thatcher era in the form of what could only be termed, with a nod to the late John McGrath, 'an exceptionally good night out' (though hardly an accessible one for the families it celebrates). Revisiting the nationwide strike by more than 150,000 workers attempting to prevent the government-ordered closure of 20 collieries, it revives the oppositions of the 1980s. Thatcher regarded the strike as a crucial test of her power and moved thousands of police officers into the collieries to protect temporary workers crossing the picket lines. The strike failed to halt pit closures, and the coal-mining regions have since become the most poverty-stricken areas of Britain. Thatcher's part in the destruction of once-thriving communities remains a bitter memory for many.

Billy Elliot opposes Thatcherite attitudes towards the working class in its songs of protest and through characterization of the miners themselves. The men who maintain the strike through months of hardship only to go down in defeat are honest, hardworking husbands, fathers, and sons. The bulk and unalloyed masculinity of their presence on stage suggests that they provide the essential foundation of their nation in terms of both the work they do and the character they exemplify. The more personal views of the miners in Billy's father and his older brother Tony do not gloss over their imperfections, but remain consistently sympathetic. Their anger spills over into violence at times, but they are provoked by the actions of an arrogant power. Their acceptance and support of Billy in spite of initial dismay at the idea of his becoming a dancer testifies to community solidarity strong enough to overcome deep-seated prejudice.

The two types of conflict – the strike and Billy's difference from his family – that form the basis of the action play out in very different ways. In terms of the public conflict, the play offers clear-cut divisions and a clear division of sympathy. Thatcher's representatives, shown as policemen from similar backgrounds but a different region from the miners, taunt the suffering strikers. The defiance of the miners, who are already losing their way of life, shows an edge of sadness foreshadowing their irreversible loss. In the personal conflict between collective consciousness and individuality, however, both sides receive sympathetic treatment. The celebration of the miners' collective unity, especially in the moments of song, expresses not only nostalgia for an earlier time, but also optimism for the future. Unity does not turn to exclusion when one of their own seeks to move beyond its boundaries; instead, it bridges the distance between their lives and the life to which Billy aspires. The miners' pride similarly strengthens them without displacing other values, as can be seen when Billy's father sacrifices his pride to obtain the money necessary for his son's audition. The individualism at the basis of Billy's ambition to be a dancer stands in marked contrast to the collective consciousness of the miners; however, Billy's is not the selfish individuality associated with the 1980s, but rather the necessarily single-minded pursuit of a personal passion. Billy's individualism, moreover, grows out of a rare physical and expressive gift, rather than a calculating ambition.

The show concludes, through the magic inherent in the musical theatre form and the understanding of director Stephen Daldry that the story has the qualities of a 'myth or fairy tale' (Programme 3), with a conjunction of Billy's individual passion and the miners' passion for their community. Billy flies ecstatically, while the miners stolidly go down into the earth, but in their separation they express a unity of spirit. The state that Billy enters when he dances is, in the words of his song 'Electricity', 'like forgetting, losing who you are / And at the same time, something makes you whole' (Programme 18). As a dancer, Billy transcends himself to become his world. The miners' credo of acting 'as one' (*ibid.*) assures Billy of continuing connection even though he is leaving the village to pursue a life none of the men would have foreseen or wanted for him. Through dance, Billy has found a way to overcome his grief at the loss of his mother. His example suggests that the miners will eventually overcome the loss of their occupation.

Post-Thatcher dramatizations of working-class life and culture acknowledge the end of many traditional industries at the same time that they demonstrate a decisive break between the predictable patterns of the past and the uncertainties of the present situation. These plays portray the past in gentle and affirming ways, even while offering vivid examples of the difficulties of jobs that carried a high element of risk and social roles that severely limited choice. They show the present through characters who have lost a sense of location in terms of their economic and social roles. Unsure who they are in the new social and economic order, these characters express frustration and confusion, sometimes turning to violence. The often-enigmatic presence of a child in the final image of these plays signifies the unknown future of the families and communities at their centre.

The transition to marginality in post-Thatcher Britain

Though *Billy Elliot* seems to hold out the possibility of an optimistic future in which the wounds inflicted upon the working class by Thatcherism will be healed,[1] a more common view makes visible the formation of a permanent underclass. A group of political plays show working-class communities forced into idleness and increasingly marginalized as they lose not only economic power but also the social and political power inherent in productivity. Such plays attempt to show the effects of joblessness and poverty on both the Thatcher and the post-Thatcher generations. Their view of young people focuses on the point in youth when the doorway to future prospects opens or closes. The notion of a doorway points to the political position of the plays, as they locate this doorway and control of it within social and political choices that may be changed. These works show the pitiable plight of many young people in contemporary Britain, but they use pity to mobilize the social conscience and pose questions about access to the education, jobs, and social services that could give such young people a better chance at a productive future.

The paired plays *Rita, Sue and Bob Too/A State Affair* (2000) take a look at the deterioration of life on a council estate over the 18-year period between the writing of *Rita, Sue and Bob Too* in 1982 and the writing of *A State Affair* in 2000. These two plays were brought together and directed by Max Stafford-Clark for Out of Joint and co-produced by the Liverpool Everyman and Playhouse and the Soho Theatre companies. *Rita, Sue and Bob Too*, by the then 21-year-old Andrea Dunbar, presents an arresting picture of life among working-class young people living on a Bradford Estate at the beginning of the Thatcher era. Dunbar, who found a dramatic voice through the Young Writers' Programme at the Royal Court, died in 1990, at the age of 29. In reviving the 1982 play, which he had directed while serving as artistic director of the Royal Court, Max Stafford-Clark, along with writer Robin Soans and a group of actors, researched contemporary housing estates in northern England. Revisiting the Buttershaw Estate where Andrea Dunbar had lived, they found two major changes: first, a heroin epidemic among the estate's young people, and second, a major rebuilding programme improving housing and security for many of its residents. These changes indicate a mixture of hopeless personal destructiveness and hopeful effort towards reconstruction within the community.

Rita, Sue and Bob Too begins with a sexually explicit and farcical scene set in the back seat of a car. Teenagers Rita and Sue take turns having sex with Bob, who is supposed to be taking them home after their babysitting job for his children. Subsequent scenes show Bob, a thirtyish contractor, meeting the girls clandestinely and arguing with his wife Michelle over her suspicion that he is having an affair. She confides her suspicions to the girls, who continue coming to the house to babysit, but does not think Bob will leave her. In bits and pieces, the girls reveal their outlook on life. Neither girl cares about school, but Rita would like to go to London and get a job with the police. Both admire the materially better life of Michelle and Bob. Sue feels no guilt about the affair with Bob, but Rita feels guilty about betraying Michelle. While neither girl communicates pleasure in sex, both eagerly take their turn with Bob.

The affair contributes to disintegration of the social bonds in their community. The girls' friendship breaks down in competition for Bob. Bob begins to have trouble getting work and experiences impotence.

Word gets out about the affair, and family quarrels ensue. Michelle leaves Bob, taking their children. Rita discovers she is pregnant and moves in with Bob. A couple of years later, at a chance meeting in the pub, Sue and Michelle talk about the break-up of Michelle's marriage and her life as a single mother. Sue's mother joins them and takes Michelle's side in condemning Rita for the affair, but Sue does not blame her. At the end, Michelle and Sue's mother remain in the pub, commiserating with each other over the trouble caused by men. *Rita, Sue and Bob Too* thus places women at the centre of working-class life, showing them as strong and pragmatic while men seem less stable and more vulnerable to changing conditions. Parents lack the social or personal power to prevent teenage daughters from sacrificing their future for the sake of present excitement. Dysfunction within families contributes to the breakdown of community and the formation of social problems.

A *State Affair*, despite its title, does not emphasize the wider context of economic and social change, but rather focuses on the perilous situation of some of the estate's young people in the post-Thatcher era. The play consists of monologues taken from interviews with people living in and near Bradford in 2000. In overlapping speeches, they tell stories of parents who abandoned or neglected them, parents whose illness and disability placed extraordinary demands on them, and parents who beat or sexually abused them. Their lives have encompassed a wide spectrum of disruptive and destructive behaviour: leaving school at 16, gang fights, early sex, promiscuity and prostitution, early pregnancy, stealing, drug use, suicide attempts. Some have been incarcerated or hospitalized. Drug addiction has caused Marie to lose custody of her child and brought Andy a crippling injury and then a horrible death. Paul chillingly evokes the extremes of addiction by stating that he would cold-bloodedly kill a child to obtain his drug. Two volunteers describe their work with these young people, the ways they have sometimes been able to help, and the overwhelming demand for help. At the end of the play Andrea Dunbar's daughter Lorraine describes the loss of community and blames drugs for the loss. She comments, 'If my mum wrote the play now, Rita and Sue would be smackheads... on crack as well... and working the red-light district' (133). She reveals anger at her mother, who had considered aborting her, and speaks of her 'outcast' status as the offspring of Dunbar's relationship with a Pakistani man. She tells about the drug habit she has overcome, and her efforts to take care of her children.

Very different moods predominate in the two plays. In *Rita, Sue and Bob Too* the mood of teenage rebellion predominates. The girls test limits and discover freedom in the short period of late adolescence before they must accept the burdens of adulthood. Their rebellion occurs against the backdrop of a relatively stable community that is harmed but not destroyed by the betrayals, arguments, and crises at the heart of the play's action. In *A State Affair* the community seems to have failed. The young people express desperation, understanding that their mistakes might not be reversible and that they might not survive their adolescence. Parents seem to be absent or part of the problem. Community institutions, such as the schools, remain marginal in their influence. Overworked volunteers labour heroically to stem the enormous tide of teenage disorder and despair. One of the volunteers offers his concept of hope:

> You get two chances... one is given you as a kid by the state, and it's quite easy to turn that down. But you do get another chance later, and you have to be ready for it when it comes along (106).

Lorraine's concluding monologue puts this sense of hope into a realistic context, as she observes that some people on the estate are 'getting their lives together with a lot of courage and determination' while others are 'going down a big steep hill, into a big black hole' (134). This pair of plays offers no solutions but provokes thought about whether or how such communities can be regenerated and how a nation should address this continuing destruction within and of its people.

In *Herons* (2001), Simon Stephens offers another poignant view of adolescents struggling to survive in a blighted urban wilderness among the new underclass. The play begins in a tranquil mood, with a setting beside an old lock where the Limehouse Cut meets the River Lee in East London. In this grimy area amid the empty relics of an industrial past, 14-year-old Billy has found a spot of solitude where nature seems to be reclaiming the city. Here he enjoys the play of light on the water, catches small fish, and watches for wild herons. The liminality of this spot, natural and man-made, part of the community but isolated from its social structures such as the police and school officials, signifies the transitional state of the adolescents who frequent it. They include, besides Billy, a group of three 15-year-old boys led by the coldly cruel Scott, and 13-year-old Adele, who is known as Scott's girlfriend but who makes overtures of friendship to Billy.

Billy has survived domestic violence and parental failure. He fled from an alcoholic and abusive mother to live with his father, who is jobless and depressed. Now Billy must endure bullying and threats from Scott and his friends. Scott justifies this abuse by blaming Billy's father for his older brother's conviction and imprisonment for the murder of a young girl. Billy's father discovered the victim near this secluded spot. The one positive element in Billy's troubled life is the growing friendship with Adele. Drawn together by sympathy for the murdered girl, Billy and Adele use the freedom of the outdoor space for quiet talks. Billy reveals to Adele the journal he keeps to try to make sense of his experiences. Adele tells Billy of her fear 'that I've got nothing to look forward to' (52). Billy has just two modest ambitions: 'I want to go out to the sea. Into the ocean. With my dad. I want to see dolphins swimming, real dolphins in the ocean. And I want to be able to ride on a roller coaster. A big fucking proper one' (55).

Scott's abuse of Billy escalates, and his bullying extends to Adele when he learns of her friendship with Billy. Strengthened by the belief that after death one will go to heaven if deserving of it, Billy staunchly stands up to Scott. When Scott intensifies the violence, beating and sodomizing him, Billy understands that he must somehow stop Scott or be killed. With no parent or authority to ask for help, he uses the one weapon he has – a gun kept by his father – and turns the tables on his tormentor.

Tranquillity returns in the final scene, but with it a painful sense that Billy's coming of age has turned tragic. He has, despite his stressful childhood, done well in school, pursued his hobby of fishing, taken care of his father, and kept a necessary distance from his mother. He has tried to make sense of his life and has employed a moral framework, condemning as 'despicable' the behaviour of those who 'dick around and act like tossers' at school (32) and pollute the canal with litter. These strengths, however, do not safeguard him from violence. Before confronting Scott with the gun, he rips up his journal, signalling an abandonment of rationality. Whether or not he is responsible for Scott's death, which occurs offstage, he becomes implicated in the violence around him. He confesses that when he held the gun to Scott's head and saw the frightened bully sobbing and pleading, he realized, for the first time, the way 'everything is just joined up' (83).

After he has resorted to violence, Billy ceases to record his experiences. He says he would like to have no memory, because everything he remembers frightens him. Billy states that he will go away, to a town by the sea, as he has wanted to do, but now this assertion carries a ring of the never-to-be-fulfilled fantasy in a young life with no remembered past or anticipated future. In the final moments, father and son agree to go home, but remain sitting, immobile in the endless present. Billy's attempt at self-preservation has moved him closer to the one thing he most fears, becoming 'broken up' like his father (69). From a generational standpoint, this play shows a young man with intelligence and ideals who longs to set things right in his corner of the world but finds himself trapped in an impossible situation created by the neglect and cruelty around him.

The plays that confront theatre audiences with Britain's underclass reveal serious social problems that resist simple political solutions from either side of the ideological divide. They highlight young people whose vulnerability to addiction, violence, and extreme dysfunction has been intensified by the absence of supportive families and communities. The plays do not point to specific solutions to the problems they make visible. Instead, they simply ask of audiences that they not ignore or devalue the individuals who, despite the failures that characterize their personal histories and the deprivation of their social environments, are struggling to make moral choices and construct meaning in their lives.

Rethinking victimization

The in-yer-face plays that presented violence without moral condemnation necessarily presented its victims without pity. Responses have taken up the question of victimization, especially in terms of morality. These plays do not invoke a simple oppressor–victim opposition, but consider the complexities involved in both positions. Each play places family dynamics at the centre of the action and uses family relationships to ask questions about power and morality. The forward motion of these plays stands in contrast to the stasis or backward-looking quality evident in many contemporary plays. Though all the plays considered in this section – works by Claire Dowie, Enda Walsh, Bryony Lavery, Rona Munro, and Debbie Tucker Green – present characters dealing with trauma and pain, most demonstrate a strong orientation towards the future. Significantly, concerns about victimization occur most clearly and often in plays written by women – in most cases, women whose previous plays express feminist ideas. While these plays contest the anti-feminist bias evident in many of the in-yer-face plays, they do not comprise or call for a resurgence of the feminist movement. They move back from the larger narratives of feminism to examine specific, individual lives, but have not closed these stories off from incorporation into feminist narratives of the future.

Easy Access (for the Boys) by Claire Dowie, first performed at Drill Hall in 1998, uses intersecting video sequences and live staging to examine the victimization of Michael, a young male prostitute who was sexually abused by his father in childhood. Michael is making a video diary, which is being filmed by his friend Gary. For the video, Michael interviews his father, asking his opinion of his son's occupation. In the live action, Gary, who was abused by his own father, questions Michael's lack of anger and urges confrontation. Michael, however, insists that he never suffered and actually loves his father. Video and live sequences show Michael moving in with Matt, a divorced man with a young daughter. Michael fantasizes about his father when he has sex with Matt, and expresses an almost reverent attitude towards the innocence of Matt's daughter.

Seeking respite from both the live-in relationship and the life of a prostitute, Michael returns home, to his father and the hippie-style bar he operates. He finds, to his alarm, that his father now has a live-in companion with a young son. Michael attempts to protect the boy, both from a desire to guard the child's innocence and from his need to feel that what he shared with his father was unique. He tells Gary about the situation, and Gary insists that he warn the boy's mother about his father's history of abuse. Michael prepares to do so, but his father anticipates and blocks the action. Having stolen Michael's video, he has edited it to portray Michael as a dangerous paedophile, and is prepared to blackmail his son with it.

Through the issue of abuse, Dowie explores the theme of control. Michael had control of his body taken from him as a child and has since struggled to establish a sense of autonomy. His use of his body to make money, his claim to love his father, and the video about his life all represent attempts to assert control. These strategies prove futile, as does Michael's attempt to protect the boy who now sleeps in his old room. Neither the confrontational strategy suggested by Gary nor the more immediately urgent one of revealing his father's abuse diminish the father's control. Through showing Michael's confusion, his inability to maintain autonomy or intimacy, and his sexual arousal when thinking of young children, the play reveals the internal struggles of a victim. It presents only the exterior of the abuser, Michael's father, but emphasizes his resourcefulness in maintaining power. With a veneer of harmlessness in his image as an aging hippie fixated on the music of Bob Dylan, Michael's father presents a chilling view of hypocrisy and danger. Just as the abuse has rendered Michael powerless to overcome victimization by his father, the play leaves its audience profoundly disturbed but unsure how to protect children from such predators or alter entrenched power differentials.

In *Bedbound*, first produced at the Dublin Theatre Festival in 2000, Enda Walsh charts the way in which an oppressor and a victim end up in exactly the same place. Throughout the play an ill-looking and soiled father and a filthy daughter crippled by a twisted back are trapped in a small, dirty bed in a boxed-in space. The absurdist style of the play emphasizes their isolation from the world, the sameness of their situation, and the difficulty of communication between them. The father has condemned both his daughter and himself to this prison through his attempt to succeed in the enterprise culture. Fragmented monologues reveal that he is a failed furniture magnate who started as a stock boy, engineered the death of the shop's owner, and then took control of the shop himself. He expanded the company into a chain of furniture shops, first in Cork City and then in Dublin. He set daunting goals for himself and succeeded for a time, meanwhile neglecting his family and treating his employees like slaves. As the stakes got higher and competition increased, he became disoriented and frantic, despotically barking out orders to his beaten-down assistant, Dan Dan. His Dublin shops failed just as he became aware that his daughter was suffering from a crippling disease. When his wife sickened and died and his daughter's body twisted, he obsessively built walls inside their home, confining them inside the innermost space. He too has been trapped in the boxed-in space since the day he flew into a rage and murdered Dan Dan for accidentally dropping a costly piece of furniture. In the silence, the father is filled with fear and the daughter with hatred of him. Their only distractions consist of repeating the story of the failed furniture empire and reading a formulaic romance novel that the daughter keeps as a memento of her mother.

The father possessed the individual drive and ruthlessness to succeed in the market economy. He started with a plan for success and was not deterred by moral considerations, growing more powerful and more monstrous through each act of victimizing others. Inevitably he destroyed those on whom he most depended, and now finds himself powerless to do anything except retell the stories of the past. The daughter, crippled and confined, has not been allowed to grow into maturity and independence. She can only rage at her father, cherish tender memories of her mother, and obsessively read the opening chapter of the romance novel. Neither the father's masculine script of business success nor the mother's feminine script of courtship and marriage has provided a happy ending. The daughter understands her victimization but cannot act against the father, who is her only companion and protector from the panic that overwhelms them when they think about their entombment. This play, in contrast to others that re-examine victimization, does not have a strong future orientation. It does, however, end with the father and daughter going to sleep, implying a natural cycle of slumber to be followed perhaps by awakening and a chance to renew the life that has almost left them.

Bryony Lavery's *Frozen*, which was first performed in 1998 at the Birmingham Rep and revised for its 2002 production at the National, explores the themes of loss and responsibility. Three characters provide different angles of vision on the kidnapping, sexual assault, and murder of a young girl. Nancy, the mother who must endure the nightmare of ten-year-old Rhona's disappearance after she has sent her on a routine errand to her grandmother's house, goes through many stages of change. Ralph, the paedophile and serial killer who commits the murder, remains at large for 20 years and then maintains a self-protective distance from others in prison. Agnetha, an American psychologist who studies the behaviour of criminals, comes to England in flight from guilt and loss after her closest colleague and one-time lover is killed suddenly and randomly in a road accident.

Nancy, speaking chiefly in monologues that create the sense of a journal, reveals the devastation and near destruction caused by the loss of her daughter. During the 20 years it takes the police to find the body of the murdered girl, Nancy channels her fear and grief into an organization for the families of missing children, but when her daughter turns out to have been dead all this time, the loss of hope brings forth inexpressible rage. Victimization changes and eventually ends Nancy's marriage. It also affects her older daughter Ingrid, who does not come to terms with the loss of her sister and the alteration of her family until she temporarily relocates to India and studies Eastern mysticism. With Ingrid's guidance, Nancy finds a way of mourning Rhona and, after considerable hesitation, meets the murderer and offers him her forgiveness. These forms of closure allow Nancy to begin rebuilding her life.

Ralph has abused and murdered seven girls, burying their bodies in a garden shed and long evading apprehension. Though his sessions with Agnetha hint at a tortured childhood and elicit her sympathy, he remains unmoved by empathy towards his victims until Nancy's visit. When he claims not to have frightened Rhona or hurt her, Nancy calmly assures him that her daughter must have been frightened and was hurt. When he reveals to her the trauma of his childhood, she compares the pain he felt as a child to that experienced by his victim. In response to Nancy's forgiveness, Ralph begins for the first time to understand and feel remorse. Their encounter thaws Ralph's frozen emotions, but he cannot cope with his painful feelings and hangs himself in his cell.

Agnetha plays out her own struggle with loss while posing the question of her academic paper, 'Serial Killing . . . a forgivable act?' (18) Locked in a quandary of anger and guilt over a sexual incident with her deceased colleague, who was married to her best friend, she agonizes over losing both relationships, but eventually reconnects with her former best friend through a simple telephone call. Agnetha bases her research on the idea that serial killers 'are driven by forces beyond their control' (78), and that their crimes should be viewed as symptoms rather than sins. Nancy's extension of goodness to Ralph, and his transformation through that goodness, challenges Agnetha's dismissal of concepts like good and evil, or sin and salvation. While she is not prepared to abandon the central thesis of her work, Agnetha does face the necessity of exploring this thesis rather than advocating it.

Control again proves a central theme. Though lacking public power, Ralph has imposed a kind of control over the lives of his victim's family, a control maintained for many years following the murder. Agnetha seeks to control understanding of the serial killer through the terms of clinical psychology. When subjected to the loss of her daughter, Nancy attempts to regain some sense of power by joining with others in a campaign to increase awareness of missing children. The play consistently and clearly demonstrates, however, that the sense of control is deceptive and that actual control is elusive. It suggests that power lies in changing oneself and moving out of frozen patterns of thinking, rather than attempting to exert power over external elements.

Without minimizing or romanticizing loss and its associated pain and disruption, *Frozen* asserts that new life can arise out of such a terrible event as the murder of a child. The emotional, intellectual, and social transformations at the heart of the play offer hope for personal and societal transformation. The survival and resurgence of what Ingrid calls 'The Life Force' (90) and its power to inspire new beginnings even after a catastrophe provide an argument against abandoning struggles to understand and improve the lives of individuals and societies. Significantly, it is not ideology but spirit that brings the transformation. Nancy's forgiveness, rather than Agnetha's clinical neutrality, opens Ralph to understanding the suffering of others and feeling contrition. Also important is the surviving daughter's leadership in the process of healing and going on. While her mother is immobilized by loss, Ingrid searches for a means of reconciliation and renewal. She becomes her mother's guide in a realm of unfamiliar ideas, strengthening her and reviving their relationship.

Abuse pervades family dynamics in *Born Bad* by Debbie Tucker Green, first produced by the Hampstead Theatre in 2003. Chairs on stage, beginning with a single chair and increasing to a circle of five as additional members are brought into the confrontation, represent the configurations within a 'blood-related black family' (2) of six. The gospel hymn music that opens the play and is hummed at times throughout suggests that religion is central in this family. Dawta propels the family into dialogue. Confronting her father, she demands that he 'say it' (3). He remains silent. She then rages at her mother, screaming epithets as the mother pleads for respect. Dawta informs the family that her father sexually molested her, and her mother sent her to him. Sister 1 supports Dawta and offers precise memories that support Dawta's accusations. Sister 1 believes Dawta was chosen for sex with their father because of her 'gift of strength' (11).

The remaining two siblings differ and disagree with Dawta. Sister 2 accuses Dawta of lying and exaggerating. Brother reveals that he, too, was abused by the father, but since the mother was not involved, he has no anger towards her. Their father had told Brother he was the only one, and learning that his sister was abused robs him of the sense of closeness that rewarded his compliance. Brother also reveals to Dawta that he made Sister 1 his confidante from an early stage. Dawta feels betrayed by the fact that Sister 1 knew both siblings were molested, but never communicated about it. Without the support of her siblings, Dawta confronts her mother, and the mother retaliates with the accusation, 'you was born bad' (31). Brother attempts to confront their father, but he maintains silence. Sister 2 still refuses to believe her siblings were abused. Neither Dawta nor Brother gains satisfaction from the confrontation.

The play simply and starkly sketches the power relations within the family. The father's silence and immobility signal his privilege. The use of single chairs and the interactions that progressively separate each family member from the other indicate how isolation serves to maintain the father's power. Words and gestures that seemingly should draw the siblings – especially the two who experienced the abuse – and their mother together instead alienate them from one another. Although their oppression has a common source, each of them experienced it differently, and these differences prevent them from creating a consistent narrative of the abuse and forming a powerful alliance against its perpetrator. Without this support, the victims lack the possibility of justice. The changing configuration of chairs has one too few, at the end, to accommodate the family. The siblings negotiate tensely over whether and how to sit, and Dawta realizes that she can protect her siblings and remain in the family circle only by sitting between her father's legs. She does so, humming the gospel hymn as protection. When she chooses her father instead of her mother, the stubbornly silent father finally speaks, blaming his wife for choosing the wrong one.

The family of *Born Bad*, though specific in its race and Afro-Caribbean dialect, exemplifies very basic and broad patterns in the operation of male privilege. The general, even metaphoric, quality of the play is emphasized through its minimal, non-realistic staging and the poetic, almost ritualistic, rhythms of its dialogue. Once formed, the circle of chairs never opens to permit the characters to exit the situation or to allow the perspectives of the audience to enter their private world. This forces the audience to see the family, and the system it represents, as a discrete unit. No outsider can change the unit or create a way out for an individual within it. The suffering evident within the circle calls out for relief and justice, but the absence of a larger frame of reference isolates the characters and denies them the power of alliance with or refuge in any structure outside the family.

Iron by Scottish playwright Rona Munro, which was first performed at Edinburgh's Traverse Theatre in 2002, shows a mother and daughter who try to connect across the barrier created by an act of family violence that has separated them for 15 years. Twenty-five-year-old Josie has no memory of her mother, Fay, because she has been incarcerated since she murdered Josie's father. At the beginning of the play she has traced her mother and come to visit. Divorced and lacking any close associations, Josie has taken a job in the city where her mother is imprisoned. She hopes to learn about her childhood, which she has blocked from memory. Fay reluctantly and with obvious anguish relates stories featuring a bright kitchen, a well-tended garden, an affectionate father, and a wilful 7-year-old determined to get her ears pierced. Fay also describes her own wild and impulsive behaviour as a young woman and her passionate love for her husband.

Over successive visits, simultaneous staging of Fay's cell and the more open visiting area emphasizes the contrasts between mother and daughter. Fay shows nervous shaking and other indications of emotional instability, while Josie appears calm and self-possessed. Josie brings fruit baskets to her mother, while her mother would prefer cigarettes. Fay urges Josie to liven up her prim appearance, go out and have fun, and bring Fay stories of her social activities so that she can enjoy them vicariously. The serious and introverted Josie instead does research into her mother's case, contacts a solicitor, and makes plans for an appeal. She learns that her mother refused to testify about the murder at her trial and becomes certain that her father subjected her mother to violence.

Munro uses the archetype of a woman pushed over the edge by a violent man to draw the audience into Josie's theory of her mother's victimization, but dashes that theory in a climactic scene. Fay finally reveals that her own intemperate nature caused both her extravagant love for her husband and the anger that provoked her to kill him after a quarrel. Her continuing love prevents her from indulging Josie's theory and betraying the memories that are all she has of him. Feeling that Josie must choose between them, Fay urges her to 'remember your Dad and go away from here and never come back' (95). The aftermath of the climactic scene offers a glimpse of both women. Josie, who has started a new job in the city where she grew up, has adopted the more sensual appearance that her mother had urged, and in conversation she relates a warm memory of her father. Fay, chatting with a prison guard whose child is about to begin school, reflects on the necessity of letting go of one's children. Her hope that Josie will be happy is so strong that she 'daren't think about it' (99). The choice Fay made to free her daughter with the truth, rather than keep her close by withholding it, becomes clear.

The play's title reflects both the strength of the women at its centre and the conditions of their life. The imprisoned Fay is separated from the world and, most importantly, from her daughter by iron bars. The guards and the ubiquitous paperwork provide a realistic visual referent for prison's restrictions. Fay never questions the justice of her imprisonment, the conditions of her life, or the relationship that caused her fatal explosion of anger. In the face of everything but her daughter's well-being, she shows an iron-like endurance. For her daughter she wants a life of freedom and happiness. Josie has locked herself behind a hard façade of professional competence and a strictly businesslike appearance. Despite her limitations, Fay still has the power of choice, and she realizes she can give Josie either herself as a victim or her father as an unblemished memory. Giving herself as victim would promote continued contact but at the cost of Josie's freedom. Therefore, she sets Josie free with an image of a loving father that she hopes will sustain her through a lifetime of happiness. The play thus offers an unexpected view of women's strength, the understanding that can arise between generations that are separated but nevertheless linked by shared experiences, and the potential for overcoming victimization and moving to a new sense of freedom.

These plays re-examine victimization in the context of the woman-centred consciousness that formed an important element of the feminist movement. They highlight power disparities and make visible types of violation often referenced in feminist analysis; yet, they do not attempt to construct paradigms for feminist thought or action. In fact, they resist cultural or societal explanations and solutions for these cases exemplifying archetypical patterns of violation and control. The plays seem to suggest an individual rather than a categorical approach to understanding victimization, focusing on personal as much as social factors in the events they portray. At the same time, they offer a moral perspective that differs from and contests the amoral tone of the in-yer-face plays. While not suggesting the revival of a movement or collective consciousness, they demonstrate the need for allying with others who share or understand the situation to overcome victimization. Most important, they offer possibilities for moving beyond the static state of victimization to re-engage with the multidimensional activity of life.

Summary

The responses to the in-yer-face plays contest the Thatcherite attitude that there is no such thing as society. They make visible societal formations such as ethnicity, race, and class that structure identity and choice, providing an alternate perspective on the sense of emptiness, absence of definition, and meaninglessness of choice evident in the in-yer-face plays. The responses take up some of the unfinished business of the 1970s movements for social change, but not with the expectation of reviving collective action. Instead, they attempt to engage the conscience of the individual, based on a humanistic code of morality, attempting to reopen a space where social issues can be discussed. They address what Sierz has called the 'crisis of liberal imagination' (20) by returning to the wellspring of liberalism, the value of each individual. They assert the value of economically unproductive and dependent individuals marginalized by Thatcher's policies and denigrated in Thatcherite rhetoric.

The responses use two common elements to make visible the value of the individual. The first is history: images highlighting social and economic contributions in the past reinstate the dignity of individuals who have been rendered unproductive and marginal by recent economic changes. The second is generational: placing threatened individuals in the context of family and movement towards the future lays the groundwork for identifying with their situation. While acknowledging the threats and even the sense of catastrophe of the in-yer-face plays, the responses find in fundamental human experiences and forms of social organization potential strength and hope for the future. A multidimensional generational dialogue continues in British drama, extending beyond the works considered in this chapter and book. Indeed, elements of it appear in many plays since 2000 addressing a wide range of issues.

4
Systems of Power

The debates about power underlying both the in-yer-face plays and the intergenerational responses have renewed political consciousness and brought fresh examinations of the structures and issues of political life in contemporary Britain. The in-yer-face plays exult in moments of personal assertion but question or deny the capacity of individuals to engage in sustained or concerted action to claim collective power. They represent the post-Thatcher generation as disabled and stalled in an unending present, lacking both knowledge and belief and unable to connect with the past or the future. The responses open the angle of vision to include both past and future, while attempting to recover a basic moral framework and a sense of social momentum. They suggest the potential for collective social power by locating strength and honour in particular identities. They also create awareness of society's collective nature by providing specific contexts for the post-Thatcher mood of futility and by gesturing towards future possibilities.

Analytical approaches to questions about political participation and power have led to plays that examine forms of power in socio-historical contexts. Society is structured through systems of power, some of which operate almost invisibly. Political theatre may aim at exposing these systems of power by bringing them to conscious attention. Brechtian techniques, for example, attempt to make visible structures of control that have been rendered invisible through acceptance as natural or inevitable. Exposed structures may be subjected to critique. Theatre need not employ Brechtian style, however, to call attention to forms of power and analyse their operation historically or in the present. Post-Thatcher playwrights use a variety of styles and techniques to defamiliarize systems of power that create the current political climate. The forms of power exposed in post-Thatcher political drama range from the personal and individual to collective formations that may or may not have overt political aims.

Governmental power

Concepts of national government necessarily occupy a central place in contemporary discussions of political power. As Edelman points out, the national government is the most visible aspect of the political spectacle. Mass media continuously present the spectacle of national leadership, keeping it at the forefront of consciousness. The leadership itself constructs the spectacle of its power to keep the public 'both apprehensive and hopeful' about the problems it chooses to highlight (120). As even her supporters acknowledge, Margaret Thatcher's confrontational and highly visible leadership style and forcible centralization of government raised new questions about the role of the prime minister.[1] Two post-Thatcher plays about governmental power focus on the spectacle of leadership, the production of that spectacle, and the behind-the-scenes competition over its production. Both deal with specific and actual forms of national government in their own historical contexts, but offer perspectives relevant to contemporary forms of governmental power.

Power by Nick Dear (2003), first performed at the National Theatre, shows the operation of a despotic government. Based on known figures in the reign of Louis XIV in seventeenth-century France, *Power* examines the desires and ambitions of individuals at the centre of a monarchy known for communicating power through spectacle. Louis XIV stands at a transition point in European history, battling to retain absolute power amid growing restlessness and revolutions like the one that had ended the reign and life of the English monarch Charles I in 1649. With an understated tone and compact sequence of events, the play focuses on character, showing the interaction of personalities in the contest for power. The action begins with the death, in 1661, of Cardinal Mazarin, who had been the First Minister and *de facto* monarch of France since Louis's accession at the age of four. Louis fears facing the future without Mazarin, but immediately determines to rule personally rather than through surrogates and councils. Those close to Louis compete for a share of the power he has decided not to share. Fouquet, the vastly wealthy Superintendent of Finance, wants to be named First Minister, and spends freely in his pursuit of power. Colbert, who had been Mazarin's secretary, offers Louis his services as a skilled bookkeeper, and rather than seeking an appointment to the treasury, he advises Louis to take personal control of it.

Fouquet shows a genius for spectacle, dominating every gathering with his extravagant dress and sparkling wit. With his fortune he acquires rare and beautiful things, which he displays in a uniquely designed chateau and gardens. When, despite his general popularity and his gifts and loans to the financially strapped royal family, he does not receive his desired appointment, Fouquet redoubles his efforts to stand out. He damages himself with clumsy mistakes, insulting the queen by offering her unsolicited advice, and making an unwitting attempt to seduce the young woman who had become the king's mistress. His desire for admiration leads to a fatal error: he invites the royal family to his chateau for a sumptuous dinner and entertainment that far exceeds anything they could offer in return. When the king has him arrested, Fouquet turns to the nation's only democratic institution, the Parliament of Paris. As its attorney general, he would have been entitled to trial by Parliament; however, in his quest for personal power, he had resigned from Parliament at Louis's request. The end of the play shows Fouquet imprisoned, bound, and hooded like the hunting hawk he had showed off to Louis in an early scene.

Colbert presents a stark contrast to Fouquet, shunning court gatherings, dressing inconspicuously, and refusing even a glass of wine. Colbert audits the treasury accounts, and though his dogged work uncovers only the customary creative accounting, he brings accusations against Fouquet. Understanding that Louis already envisions himself as the Sun King, he warns him that Fouquet, with his genius for display, outshines the monarch. The almost invisible Colbert gains Louis's absolute trust and complete control of the treasury, but never receives a ministerial appointment or any other form of visible power. For Colbert, power lies in numbers, and he succeeds in quietly amassing a fortune while financing Louis's spectacular displays of power.

Louis uses the structure of monarchy to take power from others and consolidate it in himself. While playing Fouquet and Colbert off against each other, he dallies briefly with his brother's English wife Henriette, but discards her in favour of her 16-year-old chambermaid, who presents no threat to his power and carries no disturbing association with the beheaded English king. He occupies his brother by having him draw up a system of etiquette reinforcing the singular power of the king, while subtly abetting his disagreements with Henriette so that they will not join together against him. When Louis moves against Fouquet, he does so in a highly visible way, arresting him publicly near his home and thus making an example of him. By the end of the play, Louis has confiscated Fouquet's wealth and copied the design of his chateau and gardens for the Palace of Versailles. He has commanded all his courtiers to live at Versailles, where, he observes with satisfaction, 'former warlords now assist me when I sit on my commode' (94).

The play accurately condenses this piece of French history. Fouquet was charged with embezzlement, tried over a period of three years in French courts, and eventually sentenced to banishment, which was tantamount to acquittal. In spite of the court's verdict, he remained in prison for the remaining 18 years of his life, because Louis, as 'the living law' (88), overruled the sentence. In the play's final scene Fouquet, alone in his dungeon, ruminates about power. He regrets his own vanity and quest for power. He considers Louis's reign 'the state as theatre, his life the drama' (94). Fouquet now understands that the spectacle of power becomes political reality only through sacrifice, and that he has provided the necessary sacrifice. His impressive spectacle of individual power marked him for elimination. Louis drew power to himself through the drama of removing this man from his visible position of power. As Louis explains to Fouquet, he arrested him in his home district to warn others, 'If I can take *you*, here – I can take anybody, anywhere' (84). Fouquet's imprisonment assures Louis that his singular preeminence as the Sun King will not be threatened by any of his courtiers.

Power functions both as a historical case study and an analysis of absolute power in the context of government. Though the play does not reach overtly for contemporary parallels, its portrayal of Louis's character and his methods of taking control reveals the conditions for dictatorship, even within a partially democratic society. Louis, a man without extraordinary gifts, was able not only to maintain control of his inherited kingdom throughout an exceptionally long reign, but also to create an extraordinary and enduring image of himself as a cultural and political leader. He accomplished this by using the ambitions of prominent people around him to ruthlessly annex their talents, wealth, and strength, while ensuring that he would dominate the spectacle created by these forms of power.

Michael Frayn's *Democracy*, first produced by the National Theatre in 2003, explores the power of a memorable political leader in the context of a modern parliamentary democracy. Based on actual individuals and events, *Democracy* centres on Willy Brandt, who served as West German Chancellor from 1969 to 1974, and Günter Guillaume, who spied for the East German government while serving as Brandt's personal assistant. *Democracy*, like *Power*, shows a government in the process of forming itself while it serves as the organizing structure of its nation. Brandt came to office at a time of crucial transition not only for East and West

Germany, but also for all Europe, as the insularity, suspicion, and hostility of the Cold War began to dissolve. In contrast to *Power*, it focuses on the unseen forces at the centre of a government.

The epic-style play unfolds with an inexorable rhythm marked by Guillaume's connective narration. Scenes unfold kaleidoscopically, often transforming through the subtlest of signals, such as an actor's shift of visual focus. The simultaneous presence of many men on stage gives a choral quality to the dialogue; even though they do not speak in unison, the characters often finish one another's sentences. Recurring phrases, such as 'clean hands' to refer to someone who has not made objectionable compromises and 'hand in the fire' to evoke an image of loyalty, reinforce the sense of a Greek-style chorus bearing witness to the momentous rise and fall of a hero. The silent attentiveness of actors who are on stage listening when not speaking underlines the importance of what is said in formal negotiations, casual conversations, public addresses, private reports, and personal reflections. The original set, a large, complex arrangement of spaces varying in size and prominence, created an almost organic conglomerate of public platforms, private offices, and other spaces housing a government. Dramatic action shows both the public face of governmental power, as in the announcement of an election outcome, and its behind-the-scenes operations, such as the closed-door meetings about political strategy.

Brandt's drama follows the form of classic tragedy. He ascends to power by means of personal charisma. The play opens with his breathtaking election victory at the head of the left-leaning Social Democratic Party – the first in 40 years for a leftist party. Initiating a campaign to normalize relations with East Germany, the Soviet Union, and Poland, Brandt visits East Germany, where crowds mob him and respond passionately to a silent gesture of hope. He plans strategy with his chief of staff, his party leader in the Bundestag, and his likely successor. Despite his party leader's opposition, he forms a coalition with the Liberals, rather than with his party's traditional allies, the Christian Democrats, and holds it together through personal strength. He renews his international campaign, again using the spectacle of gesture when he kneels before a monument to the Jews killed by Nazis in the Warsaw Ghetto. Brandt's ability to communicate inspires awed admiration and stifles disagreements about strategy. He wins approval of the normalization treaties and is re-elected ten days later by a large majority. At what seems the peak of his power, however, Brandt has begun to disintegrate. Faced with an economic recession, he succumbs to passivity and illness, and indulges in scandalous sexual affairs. His supporters compete for advantage in a transition that has begun to seem inevitable. The revelation that his personal assistant has been spying for East Germany provokes a storm that culminates in his resignation.[2]

Guillaume's progress from the periphery to the centre of Brandt's circle parallels Brandt's political rise. Reportedly an escapee from East Berlin, Guillaume becomes Brandt's liaison to the trade unions, even though he has not been associated with a trade. While Brandt dismisses East German threats, joking about their political police, East Germany has not given up spying. According to Guillaume's contact, Kretschmann, they want to know if Brandt's publicly stated intentions are his real intentions and if he is willing to 'pay the price' (16) for normalization. Through a succession of jobs that draw him progressively closer to Brandt, Guillaume becomes devoted to him. He describes himself as an 'upturned face like all the others' (89) cheering the leader and drinking in his words. 'He listens' (37) is Guillaume's explanation for giving his heart to the man he spies on. When he congratulates Brandt on his re-election, he describes their association as the best thing that has ever happened to him. Democracy also intrigues Guillaume. He finds the first-hand experience of an election exhilarating and begins to question East German policies, but nevertheless continues as its agent. Though he trusts and even loves Brandt, Guillaume remains trapped in the habits of enmity that Brandt sought to change. Guillaume's unstable position during most of the play keeps him circulating between Brandt's office and the café where he reports to his contact. Sitting immobile in prison at the end, he is finally confined to one side, expressing the greatest regret and shame for letting Brandt down.

The manner of his fall from power links Brandt inextricably with Guillaume, who lays bare Brandt's tragic flaw. Seeking trust, Brandt offered trust and made himself vulnerable to betrayal. Seeking change, he embraced unorthodox thinking, which gave him a distinct political identity and led to dramatic change, but left him without personal interest or political support to deal with mundane problems. The fall exposes the flaw in Brandt's government as well. The lack of a legitimate trade union representative in this government led by leftist intellectuals created the gap through which the spy entered. The liaison between trade unions and the inner circle has, of course, been ineffective. The spirit of reconciliation evident in the treaty and re-election does not prevent the trade unions from making demands that place Brandt in a no-win situation. Having worked actively and gained credit for passage of the treaties, Brandt takes a passive role and absorbs personal blame in this crisis.

Alternating between the public spectacle of power and the behind-the-scenes struggles over the construction of this spectacle, *Democracy* shows a governing party in action, dramatizes the personalities in play during a crucial transition, and provides contrasting viewpoints on the changes involved in this transition. The men at the centre of the government disagree about goals and strategy. Wehner, the party leader, distrusts coalitions because he associates their collapse with the rise of Nazism in the 1930s, but Brandt dodges Wehner and intensifies personal control by making an unexpected alliance. Brandt needs the coalition, and it, in turn, depends on his personal leadership. As the representative of decisive change, Brandt rallies people, but this form of unity proves effective only for the specific goal of the treaties. When he wins re-election in the wake of the treaties, the achievement splinters as Brandt's associates attribute to it different meanings. One credits Brandt's charisma, while another ascribes it to party reform. Even while they celebrate, they speculate on who will be dismissed from the inner circle. After the election, a deflated Guillaume observes, 'They've hauled the monument into place. And now they've dropped the ropes, there's nothing to keep them together' (59). Kretschmann predicts fragmentation, with 'sixty million separate selves, rolling about the ship like loose cargo in a storm' (*ibid.*).

This dramatization of a government in action does not romanticize democracy, but does reveal its strength, resiliency, and magnitude. In Willy Brandt's ascendancy to power, achievement of a historic shift in national consciousness, and fall from power, many hands are held in the fire and few stay clean. Brandt himself, the visible power figure, eventually becomes the sacrifice that allows the government to continue functioning. Guillaume's presence at the very centre of the government contests the idea of a purely democratic system, but leaves unanswered the question of whether his spying had any real effect on the course of events in the two Germanys of the time. Democracy itself not only embodies competing viewpoints, such as those between 'the old Communist and the old Wehrmacht officer', but also brings these opponents together in their desire for 'long-forgotten discipline' (26). The democratic nation survives the violation of trust and the fall of its leader, moving on to install a new leader steeped in tradition and focused on stabilization. The play testifies not only to democracy's capacity to move a nation beyond habitual patterns and foster major change, but also to its capacity to move beyond the tragic paradigm and find a leader to address the historical moment that follows momentous change. Ultimately, the play emphasizes the limits of individual power in the currents of historical change, as the play ends in suspension of action, with everyone listening in amazement to the noise of the Berlin Wall being dismantled.

The two plays about government show an interesting juxtaposition of the visible and invisible elements of state power. *Power* emphasizes the visible aspects of state power: the image-conscious monarch, lavish displays of wealth, and stark displays of coercion. It does, however, make the point that Louis's pre-eminence depends on the loyalty and work of the almost invisible Colbert. *Democracy* not only places greater emphasis on the almost invisible operations of Guillaume at the centre of Brandt's powerful circle, but also acknowledges the importance of Brandt's very visible appearances, speeches, and symbolic gestures in preparing the ground for reunification of the two Germanys. Both plays urge audiences to look beyond the outer and most visible manifestations of power in their own time as they analyse and interact with the strength, durability, and flexibility of governments.

The power of the individual

Western democracies are founded on the concept of the individual as the basic unit of both private and public life. Individualism played a central role in Thatcher's politics, which advocated self-reliance as an antidote to welfare dependence but became associated with images of unalloyed selfishness. Many post-Thatcher plays focus on political and personal aspects of individuality and individualism. Sarah Kane and Martin Crimp employ a postmodern critique of the individual, pointing to the fragmentation, discontinuity, and instability that propel and inhibit desire and action. Mark Ravenhill emphasizes the disempowerment of individuals by ubiquitous systems of marketing, while other in-yer-face playwrights show annihilation of the individual by the social environment or speculate about individual power unconstrained by legal or moral considerations. Simon Block and Patrick Marber, writing about personal relationships, inscribe relatedness even in an individual's separateness. The plays based upon generational formations place individual identity and choice in the contexts of family and community continuity. The plays discussed here focus most directly on the question of the power of individuals to achieve political goals.

Zinnie Harris turns familiar formulae around in *Further than the Furthest Thing* (2000), first produced by the Tron Theater, Glasgow, and the National Theatre. Instead of showing an apparently separate individual who is actually imbedded in a family and a particular environment, or an apparently powerful individual who is really powerless, Harris begins with an individual who seems unaware of politics, inseparable from family, and at one with her natural environment. This woman differentiates herself, attains awareness of her individual power, and goes on to assert that power against a substantial opponent. This dream-like, expressionistic play highlights the centrality of homeland in individual consciousness by focusing on the people of a remote island who are evacuated following a volcanic eruption. Though it does not specify the location, it adapts the true and arresting story of Tristan da Cunha, a small and inaccessible volcanic island in the South Atlantic. The island's 268 people were evacuated in 1961, following volcanic activity, and resettled in England. In 1963, despite opposition from the Colonial Office, some evacuees insisted on going back, and by the end of that year, all but 14 had returned to the island.

The islanders, as portrayed in the play, lead a simple life in their isolated world. They subsist on crayfish and the potatoes they cultivate in small patches, but depend on an annual supply boat to survive the winter. This crucial support disappeared during the worst of the Second World War, though the islanders did not know why the boat failed to come. In this extreme situation, they drew lots to condemn some to starvation so that the rest might survive. Mill and Bill survived and became foster parents to Francis, whose mother did not. When boat service resumed, Bill left the island for a period and returned with religion, baptizing the islanders and building a church. This close community shaped by hardships and hard choices confronts possible change when an outsider proposes building a factory to process crayfish, the island's only exportable product. Francis, who left the island to seek adventure, has returned with Hansen, the entrepreneur. When the islanders veto this project, Francis prepares to leave permanently, seeking opportunity and fleeing from the shame of rejection by Rebecca, whom he desires. At Mill's insistence that he keep Francis on the island, Bill visits Rebecca, who has been impregnated during rape by a gang of sailors. He agrees to secretly kill her child as soon as it is born if she consents to marry Francis. Bill keeps his part of the fatal bargain, but the wedding never takes place because the volcano erupts and everyone is evacuated.

The play's second act shows Mill's realization of self and development of political power in the context of her community's displacement. Settled in Southampton, the islanders work in a factory owned by Hansen. Their life in England, with regular jobs, security, and comfort, seems easier than the one they left behind. Francis adapts quickly, giving up the island's idiosyncratic pidgin in favour of standard English. The others do not adapt. Bill's longing for home, combined with his guilt, leads him to suicide. Rebecca hits a reporter who asks too many questions. Mill despises the 'puddings ... going up in a lift ... cinema ... umbrellas ... train rides and baths' of the new place and misses 'collecting sea shells ... digging on the patches' and the familiar rituals of home (118). While Bill becomes depressed and disoriented, Mill becomes purposeful and assumes a leadership role. Sensing that the refugees are not being told the truth, she shuns Hansen's offer of new houses and insists on returning to the island to assess damage. In seeking this choice, Mill develops from a simple and unsophisticated woman impressed by Hansen's card tricks into a courageous politician who triumphs over the powerful interests pitted against her. Driven to know the truth and return home if possible, Mill does not shrink from confronting cruel events in the island's history. In the end, Mill succeeds, though she pays a high price for victory, because by this time Bill is dead and Francis has become an Englishman.

Mill's achievement of individual power contrasts with the use of power by men in the play. Bill absolved his community of an old guilt through religion, but brings disaster on himself when he kills the baby, violating religious values. Hansen, who represents commercial values, wants to prevent the evacuees from returning home because he plans to profit from allowing the island to be used as a nuclear test site.[3] His values also prove unstable, however, as Mill's surprising strength and resourcefulness, combined with her humble appearance and unsophisticated speech, effects a change in him. Eventually, he not only allows the evacuees to return home but also promises 'there won't ever be a year without a boat ... even if only one of you goes back' (170). Mill's stable identity and values arise from a connection to place that she fully comprehends in exile. Realizing that her strength depends on connection to the island, she treats the campaign to return home with the same determination that enabled the islanders to survive the earlier crisis when the supply deliveries ceased. Her assertion of personal will in a situation that seems to epitomize powerlessness suggests remarkable potential for individual power grounded in a strong commitment.

Richard Bean offers a less idealistic view of individual power in *The Mentalists* (2002), produced by the National Theatre. In this play a pair of long-time friends, Morrie, a barber and amateur video maker, and Ted, an industrial manager with a dream, meet in a seedy hotel. Having discovered a book on Skinnerian operant conditioning in a disused shed, Ted wants to share this idea with the world. He promises to pay Morrie £50 to produce a video in which he presents operant conditioning as a solution to poverty, crime, and war. As the two men attempt to make the video, they hear footsteps outside the room, and then hear the door being locked from outside. Ted becomes almost hysterical with anxiety, but Morrie continues chattering unconcernedly. At length, Ted reveals that he has murdered a man with a body size similar to his own, as part of a scheme to gain his freedom by killing his family, planting the man's body, and setting the house on fire to make the crime look like a murder-suicide. As the two wait for the police, Morrie does the only thing he can do for his old friend: he gives him a shampoo and haircut.

The Mentalists sets up a contrast between two concepts of individual power. Though the extreme contrast approaches absurdity, the play remains realistic. Possessed of an *idée fixe*, Ted does not examine the weaknesses or implications of his dual plan to 'end inhumanity and injustice' (40) and gain personal freedom. His sense of individual power, based on the illusion that he has discovered 'the holy grail of how to live' (61), blinds him to the implications of murder and his proposed panacea, which would require an authoritarian system of control. Morrie, though he suspects that his friend has lost contact with reality, does not argue with him or disparage the video project; but when he learns that Ted has committed murder and plans to kill his family, he can barely voice his revulsion. By now, however, he realizes the police are outside, and he assumes the familiar role of tending Ted's hair. Morrie has no grandiose delusions about his personal power, but in the care he gives his friend he demonstrates calm loyalty and care. This play's study of individual power shows the potential for savage self-aggrandizement in an ordinary individual and offers a glimpse of the individual madness underlying authoritarian systems.

Martin McDonagh has taken the power of the individual as his primary subject. The characters of his Leenane trilogy evoke the potentials of individual power through extremes of desire, strength, weakness, and cruelty. McDonagh's most recent plays problematize individual power through bizarre clashes of individual will. *The Lieutenant of Inishmore* was first performed in 2001 at the Royal Shakespeare's Other Place in Stratford-upon-Avon. Its central character Padraic is a second lieutenant in a tiny splinter group of the paramilitary INLA. Considered a violent lunatic in his Galway village, Padraic has gone north to engage in pub bombings and vigilante actions. He is torturing a drug dealer, clearly enjoying this exercise of power, when a phone call from his father interrupts with the news that Padraic's cat Wee Tom has been taken ill. Tearful with concern for his feline friend, Padraic frees the dealer and hastens home.

In Galway, Padraic's drunken father Donny and mentally slow neighbour Davey panic over what Padraic will do if he learns that Wee Tom is dead. They have looked for a substitute black cat, but have found only an orange one, which they attempt to blacken with shoe polish in the hope of deceiving Padraic. Meanwhile, three members of the INLA are hiding out and waiting for Padraic's return. They have killed the cat to lure Padraic and assassinate him, to end his zealous persecution of drug dealers that has threatened sources of INLA financing. Davey's 16-year-old sister Mairead also awaits Padraic's return as she dreams of joining him in a paramilitary and practises her aim by shooting the eyes of cows with her air rifle.

The comedy turns on a repeated pattern of individual actions attempted, interrupted, and ultimately thwarted. Padraic shoots the orange cat and prepares to do the same to Donny and Davey. The three INLA men interrupt Padraic and take him outside for execution. Mairead, lurking nearby, shoots their eyes out with her air rifle. Padraic finishes off two of the men and is about to kill the third, when that man confesses to killing the cat, and is then taken into another room for torture. Padraic announces that he will marry Mairead, but she discovers the dead orange cat, recognizes it as her own beloved Sir Roger, and shoots Padraic in retaliation.

The play illustrates the absurdity of individual motivations, the dangerous extremes to which they may be pursued, and the futility of individual action. The exaggerated character and behaviour of Padraic and Mairead parody the heroism of Irish freedom fighters as a form of madness. The argument between the three plotters over the role of cats in the liberation of Ireland renders absurd their reverence to the principle of 'Ireland free'. The running jokes on loyalties to particular animals – Mairead cherishes cats but shoots out the eyes of cows, and Padraic's attachment to his cat does not lead to mercy for a drug dealer's spaniel – question the rationality of any type of loyalty. The continual interruptions and redirections of the action show individuals more at the mercy of one another than in charge of their fate. Finally, the surprise ending, when Wee Tom reappears, forcing Donny and Davey to conclude that the cat killed in the road must have been a similar-looking stray, reveals the limits on individual power created by inescapable inadequacies of knowledge and understanding. The cat proves, finally, the only independent character in the play.

McDonagh carries this sense of the individual as creator and victim of absurdity into *The Pillowman*, first presented at the National Theatre in 2004. Its central character, Katurian, writes absurd fables containing gruesome accounts of violence inflicted on children. He is detained and interrogated by two agents of an authoritarian regime, Tupolski and Ariel. They threaten and torture Katurian and force him to listen to the screams of his mentally handicapped brother Michal, whom they are also holding. Katurian learns that he is under suspicion because two children have recently been murdered in ways consistent with the violence in his stories, and a third child is missing. Throughout his interrogation he offers arguments for the freedom of the artist. He also reveals a personal narrative of monstrous parents who subjected his brother to deprivation and nightly tortures while showering Katurian with love in order to stimulate his imagination and writing talent. At the age of 14, Katurian murdered both parents and freed his now brain-damaged brother.

Katurian learns that he cannot separate his identity and action from that of his brother. He demands to see his brother and tries to protect him from torture, but when finally allowed to see Michal, he finds him unhurt. Discovering that the screams were engineered to pressure him leads Katurian to question any apparent reality that cannot be independently verified, and he speculates that the reported deaths of the children could be contrivances. Michal requests his favourite story about the Pillowman, a magical being who saves future suicides from a life of suffering by killing them when they are children. Then he confesses to Katurian that he has committed the murders as a means of testing the realism of the stories. Though angry at his brother, Katurian continues to comfort him, but when Michal goes to sleep, Katurian smothers him with a pillow to prevent his further suffering. Finally, Katurian confesses to all the murders – of his parents, his brother, and the three children – but does so on condition that his stories are preserved. As Ariel and Tupolski make preparations for Katurian's execution, interruptions stall and redirect the action. The two agents become more human as Ariel reveals that he was abused by his father and Tupolski exposes his own literary ambitions. The missing child is found alive, which invalidates Katurian's confession. At this point, Ariel and Tupolski agree that there is very little reason to execute Katurian but no justification for preserving his stories. The end, however, reverses this rationality, as Katurian is shot and his stories are, through Tupolski's impulsive decision, saved.

Pillowman critiques the power of the individual as represented by the artist in society. It satirizes the concept of the writer as an agent of social change by indicating that the brain-damaged Michal mimicked the action of the stories without understanding his actions. Katurian's narrative of developing as an artist through vicariously experiencing the suffering of his brother undermines the individuality of individual creativity. The story of the Pillowman points to an existential despair at the heart of creativity and sets up a conscious parallel between artistic creation and committing suicide. Katurian values his own and his brother's life solely in the context of the stories he has written. In attempting to preserve the stories, he seeks to extend their individual power beyond the boundaries of their lives. The presence of the stories in the play shows fulfilment of Katurian's desire, but the final moment of the play, narrated by the already-dead Katurian, insists that this fulfilment be seen as a random and inexplicable outcome rather than the result of personal intention or rational action.

These explorations of individual thought and action discredit the Thatcherite idea of placing social responsibility in the hands of individuals. *Further than the Furthest Thing*, grounding individual power in a mystical connection to place, gives individual desire the potential for an almost irrational resistance rather than a capacity for personal nurturing. Bean and McDonagh portray the politics of the individual as frankly irrational and antisocial. Their plays show disastrous consequences of individual attempts to formulate and act on a political philosophy, while also demonstrating the absurdity of such singular attempts to exercise power. The best thing that can be said about individual power in their plays is that the actions of different individuals, even when linked by friendship or shared political allegiances, usually cancel one another out.

Power and collective action

Collective power comes into play when individuals band together outside the formal structures of government, and sometimes in opposition to them, to advance a common cause. Advocacy and celebration of collective power characterized the counter-cultural movements of the 1970s and early 1980s, and the alternative theatre of that period. Though some plays of the pre-Thatcher period questioned collective passion and action, scepticism about the potential of collective movements has contributed to the post-Thatcher mood of disengagement. Nevertheless, playwrights and audiences remain interested in the potential for change mobilized by the convergence of people with a common passion. Two recent plays dramatize the potential and limits of collective power.

The Riot by Nick Darke (1948–2005) bases its action on an 1896 dispute between two groups of Cornish seamen. It was co-produced by the Cornwall-based Kneehigh Theatre and the National Theatre in 1999. The issue at the forefront of the struggle is freedom of religion. Fishermen in the Cornish villages of Newlyn, Mousehole, and Porthleven belong to a Christian sect that forbids Sunday work. Their income is threatened when fishermen from the north, referred to as Yorkies, begin monopolizing the market by fishing on Sunday and bringing in their catch on Monday. The Sabbath-observing fishermen, called 'buccas', plan a protest in which they will board the Yorkie boats and throw their catch overboard. They plan to keep the protest non-violent, and succeed until the fleet owner sends the word to the boats not yet in port to divert to the village of Penzance, with which the buccas have a long-standing enmity.

As the chaotic progression of events from controlled protest to open combat occurs, most of the scenes take place in the home of Bolitho, a prominent businessman and the local magistrate. Bolitho's official position and multiple financial interests place him at the centre of the local economy. In gestic scenes punctuated by sung hymns, tumbling one upon another in an array of movement and images, the collective and individual strands of the conflict find their way to his house. Representatives of local institutions include the constable requesting reinforcements for his force of two men, the owner of a Yorkie fleet demanding compensation, and the owner of the bank hoping to prevent its collapse. Meanwhile, one of Bolitho's housemaids shelters her injured Yorkie sweetheart, another steals kitchen knives to arm the buccas, and a new arrival blames Bolitho for her brother's death in Africa, because he went there after Bolitho closed the tin mine in which he had worked.

Bolitho manages to stay in control initially, but as the riot grows outside, disasters multiply inside Bolitho's house with farcical intensity. He agrees to compensate the Yorkie fleet owner so that he will not take his business elsewhere, but that decision puts him at odds with the buccas, and the new kitchen maid challenges his closing of the mine, given his financial ability to save the port. He refuses even sympathy to the distressed bank owner, who hangs himself in the house. One maid stabs another. Bolitho's mother, who had been thought dead and moved to the undertaker, comes in the door demanding breakfast. The constable arrives with the protest leader in chains, but a threatening mob has gathered. Hoping to escape the mob, Bolitho orders the constable to free the rioter, but the rioters break in and place Bolitho on trial. The trial exposes the many grievances centred on Bolitho, but the army arrives in time to save him, and afterwards he tries to make amends.

Though it skims over the outcome of the dispute, which historically the Newlyn fishermen did not win, the play contrives happy endings for individual characters highlighted in the course of events. Bolitho signs the bank over to the argumentative kitchen maid, and she accepts the marriage proposal of the fisherman who led the protest. The injured

Yorkie will marry the housemaid who sheltered him and captain one of Bolitho's boats. Bolitho rewards loyal employees and enjoys the gratitude of the community as he anticipates new business opportunities in the coming twentieth century. The collective disperses into a collection of individuals who have made some gains, though they have lost the cause that led them to take action. Those who fervently sang hymns at the beginning will now pursue individual ambitions and interests. They may soon echo the Yorkie fleet owner's opinion that 'You can't allow God to intercede in business' (18). While it begins as a celebration of collective power, *The Riot* demonstrates the collective's fragility and shows its ultimate destruction, not simply through defeat but through the selection and reward of key individuals in a classic use of co-optation. Reversion to the status quo brings a laugh rather than a gesture of loss, with a final reprise of the running joke in which Bolitho's octogenarian mother seeks a sexual partner.

Mark Ravenhill responded to, expanded upon, and revised the perspectives on history, choice, and collective power that he had introduced in *Shopping and Fucking* and elaborated on in later plays. His 2001 play, *Mother Clap's Molly House*, was commissioned by the National Theatre and first performed in the Lyttleton. Although Ravenhill is a gay playwright, *Mother Clap's Molly House* is the first of his plays with a theme based on the collective experience of homosexuals and power of queer individuals. It presents two related plot lines in alternating scenes. The first is set in an eighteenth-century molly house, a gay brothel and place of entertainment that was part of a historically documented homosexual culture of London's past (see Norton, 1992). The second is set in a contemporary London flat, where gay men party in an atmosphere of sexual display and open drug use. Doubling of roles, the mythical character Eros, and a consciously constructed parallel between the eighteenth-century Martin and the present-day Will, who want monogamous relationships but love men who are not similarly inclined, link the two contrasting worlds. The Brechtian structure of episodes with songs serves to highlight the theme of individual and collective identity.

Ravenhill's view of the pre-modern world of the molly house shows it as a place that fosters personal choice and change, even in the context of business. A sturdy and clearly bounded, if tawdry, institution, it provides a home, a livelihood, and a meeting place for sexual misfits in eighteenth-century London. It enables them to create an alternative society at odds with but occupying a niche within the surrounding social structure. Because it is a business in a nation that, long before Thatcher, allowed considerable latitude to money-making enterprises, the molly house provides a space in which these individuals can reveal their usually hidden commonality. As the song that concludes the first act states, the deity of traditional morality accommodates to Eros in the name of profit, forming a relationship that marries 'purse and arse and heart' (56).

The molly house's celebratory atmosphere emphasizes the joy of escape from a world where rigid lines separate fantasy from reality into a place where imagination rules. Nevertheless, the games and fantasies fail to connect individuals more than temporarily, because they collide with non-negotiable aspects of the real world and fail to bridge differences. During one night of partying, when Martin becomes distressed by his lover Orme's desire for a different sex partner, Mother Clap promises to reunite the couple with a new game. Blindfolding Orme and sending him out, she directs Martin in performing a mime of pregnancy, then brings Orme back to witness the 'birth' of the 'infant love' he has fathered. Imagination, however, fails to transform the wooden doll produced from beneath Martin's petticoats, and Orme refuses the role of husband or father. Similarly, when the transvestite Princess declares love to Mother Clap and offers her the choice of 'man, woman, or hermaphrodite' (77), she tells him that her 'dry old body' denies her the one thing she wants – 'life inside me' (78).

While acknowledging limits on imagination, the play celebrates its power to transform lives. Martin begins as a confused adolescent, learns the nature of his desire, overcomes his shame, enters into a romantic relationship with Orme, struggles when his faithfulness is not reciprocated, and finally goes on to a new life with the hope that his beloved will eventually join him. Princess experiments with both female and male gender identification, finding in the process a love for Tull that makes gender irrelevant. The prostitute Amy finds strength and independence in the masculine identity of Ned. Tull moves from helpless dependence to head of a successful business and then on to full autonomy when she divests herself of the business and departs with her partner Princess to pursue a new life that will take some of the transformative energy of the molly house into the world beyond it.

The contemporary scenes show gay culture emptied and exhausted by freedom, lacking collective power, and conditioned by products marketed to gay men. Devoid of imaginative possibility, Act II's present-day party contains none of the molly house's capacity for resistance, celebration, or change. Its mostly middle-aged and middle-class participants exude a mood of futility throughout a dreary orgy of partner swapping, sex toys, porno films, and videotaping. Will, the flat's owner, tries above all to protect his sofa from stains, and thus avoid permanent reminders of the party. Tom, a young working-class man new to London, counters the mood of futility by claiming a personal meaning for the party. He considers it an exciting milestone in his developing gay identity and marvels at his own transition from 'Old Me ... All scared and no sex and no drugs' to 'New Me' identified by 'Clubs. E. Shagging all sorts of blokes' (64). With no sympathetic Tull to advise and protect him, Tom finds himself ridiculed by the other partiers and soon departs, seeking an acceptance in the larger society that he has not found in a specifically gay environment. The men who remain show little pleasure in interactions dependent on drugs, alcohol, pornographic videos, and sex toys. Death and suffering, as well as sex, have become matters of habit. One man apologizes for his partner's annoying behaviour, explaining that he excuses it because of the man's HIV-positive status. Tina, a young woman who has accompanied her drug-dealer boyfriend to the flat, disrupts the party when she begins to bleed heavily from one of her many self-inflicted piercings and volunteers that she hurts herself to avoid choice and negate meaning.

In this play Ravenhill suggests the need for a collective history and identity. The metaphor of the molly house implies that the foundation of gay culture, while offering a haven for the imagination, has been commercial and escapist. Its limitations, now abundantly evident in the contemporary scenes, emphasize the need for new models of living in the larger world while maintaining distinctive identities and lifestyles. A hopeful moment occurs at the end when a molly house celebration transforms into a contemporary rave club. This moment suggests the leap out of time that unites past and present possibilities. The play's indirect and tentative politics thus make an implicit argument for historical consciousness and express hope for renewed collective energy.

Both *The Riot* and *Mother Clap's Molly House* show collective energy as a potentially empowering force but display a post-Thatcher questioning of its political effectiveness. The collective shapes and validates personal identity, giving participants a sense of power; this power, however, proves illusory in the surrounding political context. The protest in *The Riot* overwhelms its originators, leading to violence and encompassing new issues. The targets of the protest easily dismantle the collective by selectively rewarding individuals and diverting them from group concerns to personal ones. *Mother Clap's Molly House* acknowledges the social energy generated by the celebration and collective exploration of desires and identities censured by society; however, it indicates that such energy may be short-lived. Both plays show collective intent as difficult to maintain, and neither shows collective action as effective in achieving political goals.

The power of belief

Religion plays an increasingly visible role in the definition of individuals and groups within post-Thatcher Britain, as it has become a major factor in power struggles nationally and internationally. Global developments encompassing Britain often place religious and social identities in conflict. Though plays about religion were almost unknown in the latter part of the twentieth century, the new century has seen a number of plays that consider religious faith in relation to public issues. The best-known recent play with a religious setting, *Behzti (Dishonour)*, by Gupreet Kaur Bhatti, received much publicity in 2004 when its scheduled opening at the Birmingham Repertory Theatre was cancelled in response to vocal protests and threats of violence by local Sikhs. Set in a Sikh temple, *Behzti* focuses on the sexual assault of a young woman by a highly placed member of the community, and the hypocrisy and collusion surrounding this act. It examines the way in which a community's insularity protects the power and interests of its internally powerful elite. Its implied conflict between a traditional culture that limits the autonomy of women and a larger one that confers the same rights on women and men, reflects the dynamics between the state and a number of religious traditions brought into Britain from Asia and Africa. The play offended many in Birmingham's Sikh community, and became known chiefly for the controversy it generated, which raised important questions about freedom of expression and respect for religious minorities.

Contemporary plays about religious belief may analyse the power of a religious organization, as does *Behzti*, or approach faith from the standpoint of the individual, looking at the way it structures choice and meaning in a person's life. Several recent plays consider the power of religion to oppose the materialism and competitiveness of late capitalist societies by offering alternatives to marketplace values. A religious impulse impels the central incident in Helen Edmundson's *Mother Teresa Is Dead*, presented at the Royal Court in 2002. Jane has left England, abandoning her husband and five-year-old son, to work anonymously at a shelter for abandoned children in India. After weeks of not knowing her whereabouts, her husband finds her in Chennai, being cared for by an expatriate artist after an apparent mental breakdown. Jane reveals herself gradually through arguments and misunderstandings with her husband Mark and tortured changes of mind about whether to stay in India or return home. In India she has found new meaning in the ideals of the shelter's charismatic director. Work with the homeless children, however, has not assuaged the extreme guilt that compelled her journey and caused her breakdown after she attempted to exchange identity with an impoverished Indian mother. Her husband's appearance fills Jane with further guilt about abandoning him and her son, while her realization that the idealistic shelter director is also the artist's lover lessens his spiritual appeal. She decides to return home, but the artist who has been caring for Jane understands her confusion and suggests meditating before starting the journey. The play ends as the characters meditate, not revealing Jane's final decision.

This play shows one individual's attempt to translate guilt and spiritual emptiness into social action to help the poor in a developing country. As it reveals the self-centredness of Jane's gesture, it questions the potential of guilt to motivate or accomplish social change, and the capacity of spiritual belief to structure meaning in a broader sense than that of personal fulfilment. While it highlights the moral issue of poverty in developing countries, it also points to the difficulty of solving this problem. It reveals that even the shelter director Srinivas, who passionately articulates the needs of homeless children, shows uncertainty about the best approach to his nation's poverty-ridden underclass. As his artist mistress notes, Srinivas has phased through different approaches, formerly arguing at one time that big business would solve the problem and at another that art, rather than direct relief, was the best route. Edmundson adds the important point that poor people cannot be defined solely by their poverty; they exercise some degree of choice and do not necessarily want the forms of assistance offered to them by outsiders. When Jane offers the impoverished Indian woman the tokens of her own identity, including money, the woman refuses them and defends her child against what she interprets as Jane's attempt to buy him. The focus on affluent individuals troubled by poverty emphasizes the role of economic and political power in maintaining the huge gulf between rich and poor, while its suggestion of moving beyond personal guilt or action implies that new patterns of thinking must be applied to the problem.

Two recent plays that explore religion in personal terms suggest its potential in opposing the dehumanizing aspects of contemporary society. *Howard Katz* by Patrick Marber, first produced in 2001 by the National Theatre and transferred to the West End, shows an individual in crisis. An expressionistic collage of brief scenes shows Katz, a successful talent agent with a wife and son, trying to find an idea that gives his life meaning. Chatting with his parents and brother in the family barbershop, dealing with clients and co-workers at the office, doing business with television producers, visiting the zoo with his son, arguing with his wife, Katz finds power but no satisfaction. Though certain of his financial worth, he cannot feel a sense of worth and purpose that goes beyond the monetary. In rapid sequence, Katz's father dies, his wife ends their marriage, and he loses his job. Abandoning any attempt at respectability, Katz sleeps in a flophouse and eventually on the street. In a spiral of failure, he is unable to have sex with a prostitute or buy a gun on the street.

When Katz encounters an old friend who, despite unemployment and injury, maintains that 'every day's a birthday' (78), he tries to turn his life around. Attempting to speak with his wife and former employer, however, brings further rejection and estrangement. Donning a skullcap symbolizing his Jewish identity, Katz goes to a casino, where he gambles desperately in a quasi-religious ritual of turning his life over to a higher power. He loses everything, and considers it a negative verdict on his life. With a razor and bottles of rum and pills, he sits in a park and contemplates suicide. A panhandler tries to talk Katz out of suicide, but refuses his request for a hug. Alone and completely powerless, Katz begins to pray after a fashion. Half-seriously, he challenges God to send him a sign of his life's value, and is amazed to remember vividly the day his son was born. Re-experiencing the connection and meaning he felt then, Katz prays again, this time asking, 'Tell me how to live' (110).

Katz demonstrates the absence of satisfaction or meaning in career success. His wealth has enabled him to rise above his parents' modest circumstances but has not given him a sense of worth. He markets celebrities, and thus understands the emptiness of their power. As disdain for his work turns to self-disgust, he pushes family members and friends away. Unable to simply divest himself of the shell of power he inhabits, he behaves so outrageously that he is forced from it. After he has lost his position, he experiences real powerlessness. Only when he fails in a half-hearted attempt to end his life does he reach the understanding that the value of a human life lies in its connection to others through love. Remembering the birth of his son allows him to find the possibility of an individuality that does not begin with material or social power. As a result Katz experiences love that extends beyond his son to encompass all life, even the moss growing in a crack in the pavement.

Mike Leigh's *Two Thousand Years*, produced by the National in 2005, employs a classically realistic style to examine questions of identity, meaning, and faith in the context of the secularism typical of Jews in Britain. Rachel and Danny, the middle-aged couple whose comfortable

North London home provides the setting for family interactions, maintain a Jewish cultural identity but do not engage in religious practice. The family, which includes two adult children and Rachel's father, offer strong political opinions in their frequent, passionate discussions of national and world events. Their talk, even of trivial matters like traffic and parking, often turns argumentative and even angry. Both the parents and grandfather express disillusionment with causes they once held dear, from Zionism to socialism. Only the daughter Tammy, who works for Amnesty International, maintains a sense of idealistic optimism, despite her pragmatic Israeli boyfriend's reminders that her hopeful views are based on selective consideration of the 'facts on the ground' (81).

A family crisis begins when the son Josh, a man in his early thirties with a degree in mathematics but no job or regular occupation, begins to wear a skullcap, pray regularly, and show other signs of religiosity. These choices move him outside the family's secular mainstream. He adopts an appearance and diet different from the rest of the family, retreats into a brooding silence that seems judgemental, and refuses to explain his religious choices. His baffled father complains, 'it's like having a Muslim in the house' (20). Josh's grandfather derides him mercilessly. His younger sister treats him with the sibling's half-serious sarcasm but signals sympathy in her observation that Josh is 'very scared' (34).

Amid this conflict, the death of Rachel's mother brings back into the extended family her younger sister, Michelle. Having belatedly learned the news, Michelle arrives unannounced after an 11-year absence. She collapses in a histrionic display of grief that soon gives way to childish complaints of being unloved and urgent demands for a drink. Michelle's career as an investment banker and her materialist values have long separated her from the family. Now she retaliates by disparaging Rachel's work as a homemaker and taunting both her sister and father with their lack of wealth. Michelle's attacks and self-centred misery provoke chaotic arguing that resolves into hasty departures. Left alone, Michelle finds a bottle of whisky, gulps from it, and hurriedly leaves.

This intensification of the family's crisis actually solves it. Josh has observed the self-pitying alienation of his aunt in the context of his own imperfect but loving family. He has heard his mother reply to her sister's attacks by straightforwardly acknowledging that her life has meaning only through the 'people I love, and who love me' (99). His grandfather defended Josh's religious choices to Michelle as 'something that has some meaning for him – something positive' (105). He now realizes that the old man's sarcastic comments spring from disenchantment with religion, but do not negate his deep love for him.

Unobtrusively removing his skullcap, Josh rejoins his family, moving deliberately from the periphery of the gathering into its midst. The final scene shows the resumption of family rituals, with Rachel reading the *Guardian* while Danny and Josh play chess.

Two Thousand Years shows Josh's attempt to find refuge in religious ritual when political conditions fill him with fear and propel him towards despair. Religion, however, fails to protect him from conflict. His religious choices create a barrier in the family, and his aunt's loneliness makes him long for restoration of family unity. Religion provides him neither with answers nor with the strength of personal conviction his mother showed when she acknowledged her own devotion to the family. Rachel's simple statement of the meaning she has found in her choice offers Josh the hope he needs. The genuine love that he then recognizes between his puckish father and down-to-earth mother suggests an antidote to despair. Finally, when Danny characteristically tells one of his silly, religion-themed jokes, Josh seems to understand that this nonsense is his father's way of attempting to communicate a deep sense of closeness beyond language. Like *Howard Katz*, this play emphasizes human connection, rather than wealth or status, as a source of meaning, but further suggests that family provides a more important basis for meaning than religion. Both plays focus on human limitations and vulnerability, pointing to religion's potential to counter feelings of powerlessness and despair. Both, however, show individuals who find strength in Judaism's secular humanistic values rather than in its sacred texts or rituals.

Paul by Howard Brenton, which premiered at the National Theatre in 2005, approaches religion as an institution supported, as in the case of contemporary theocracies, by the power of the state. It explores religious zealotry through the life of Paul, the early Christian apostle who moved the new faith from local cult to world religion. Action takes place in a Middle East that looks modern, with camouflage-clad soldiers carrying out operations amid the rubble of destroyed buildings. Saul of Tarsus, a captain in the Temple Guards, leads the campaign to find and eliminate members of the heretical Christian sect. Under his stern leadership, the guards have become divided and weary, but Saul rallies them and offers to take the first watch. While on watch, he has one of the fainting attacks to which he is subject. He awakes to an encounter with Yeshua of Nazareth, whose peaceful demeanour, long hair, and homespun tunic create a startling contrast to the military men. Yeshua gives Saul the new name Paul and leaves him with a commitment to the faith he had previously been determined to wipe out.

Scenes of Paul awaiting execution, along with Peter, in a Roman prison frame the narrative line, which covers the period from 39 to 65 CE and follows the outlines of what is known about Paul as a historical figure. In the prison cell Paul remembers major events in his life, as Peter questions the wisdom of his far-ranging travels and conversion of Gentiles and pagans; Paul's memories range over scenes in Jerusalem, Corinth, and Rome. In one, Paul and Barnabas meet Mary Magdalene, who introduces herself as Yeshua's wife and tells variant versions of gospel stories. In a scene with Yeshua's brother James, Paul asks forgiveness for persecuting the sect and permission to take the message of Christ to a wider world. Anxious about control of his brother's message, James insists that it was intended only for Jews and aimed at reforming Judaism. Paul cannot believe that the one he has begun to call the Lord Jesus Christ aimed only at reform, believing that he came 'to change everything, to end everything' (44). James compromises, agreeing to the mission only if Paul collects money for the sect in Jerusalem wherever he goes. A later scene shows Paul in Corinth with new converts, mediating financial disputes, deciding questions of ritual, ruling on issues of sexual morality, and beautifully presenting his views, contained in the Letter to the Corinthians, on the power of selfless love.

As they wait for execution, Peter reveals to Paul the truth about his vision on the road to Damascus – a truth given substance in the realistic action of earlier scenes. Peter says that Yeshua had by chance survived his crucifixion and lived in hiding during the period of Saul's persecutions. Desperate to stop Saul's relentless campaign against them, Yeshua's followers convinced him to speak with Saul when he was alone. Peter describes Yeshua as a great teacher who straightforwardly preached kindness, but was not the saviour of mankind. He goes on to say that though he suffered poor health as a result of his ordeal, Yeshua lived on for 20 years after his encounter with Paul, with his wife Mary Magdalene. Paul repudiates Peter's revelations, and the implication that his ministry and all the conversions and new congregations have been based on a fabrication. He says, 'If Christ has not been raised our faith is pointless' (74–75).

At this critical moment the supreme representative of temporal power appears as unexpectedly as Yeshua had earlier appeared on the road to Damascus. The Emperor Nero arrives to converse with the two men he regards as effectively dead, and therefore suitable confidants for a man at the pinnacle of power. Nero scoffs at Christianity's threat to Roman religion, but does consider it a political threat. Describing Paul and Peter as 'leaders of a death cult', Nero remarks that 'death cults always give the state problems' (78). He predicts that Rome will appropriate the

Christian religion for its own purposes, relying on its priests, 'a good hierarchy of bribable gentlemen in fine robes' (79) to enforce obedience to state authority. He assures Paul and Peter that their execution will add necessary martyrs to Christianity's rather bare story. Confronted in this way, Paul declares his faith and begins to preach. Nero questions Peter, and Peter responds with the statement of faith, 'Christ died for my sins and rose from the dead' (81), making this assertion as a conscious existential choice.

Paul, as a historic individual, ascends to power through religion. Though his access to power occurs, ironically, at his moment of greatest vulnerability, Paul finds in the belief that seizes him a means of seizing upon certainty. Orienting his life completely around the new belief, he pursues converts with the same single-minded intensity that previously marked his persecutions of believers. He builds the fledgling religion into a transnational power, reinforced by a strong and well-financed organization, but remains uncorrupted by money or influence. He, as Mary Magdalene recognizes, needs religion as urgently as most men need sex. His intense need prompts his stubborn certainty that Yeshua died for the sins of the world and rose from the dead, even when Peter offers testimony invalidating those beliefs.

Neither Paul's personal power nor that of the religion he founds can contest the power of the Roman Empire. He and Peter remain in prison throughout, remembering scenes of the past rather than working actively on behalf of their religion. The sophisticated and imposing state exemplified by Nero contrasts markedly with the primitive one evidenced by Saul's roughly clad company of men in arms who fought openly in the front line of religious division. The image of the urbane young emperor, with his luxurious clothing and casual manner, masks the military might and systematic cruelty that sustains his power. Nero toys with Peter and Paul because it gives him pleasure. The state will execute them, but will most effectively control religion by patronizing and corrupting its leaders. This portrait of the beginnings of a powerful religion shows it, paradoxically as both a vehicle of rebellion against the state and a compliant agent of the state, and as both an arbitrary framework to assert certainty and a beautiful if uncertain route to understanding profound truths.

Clearly, these plays frame religion within secular assumptions, viewing it as a form of collective power rather than as an instrument of divine truth. As such, the prevailing view of religion is much like that of informal collectives: neither fills the contemporary political vacuum nor provides a substitute for discredited forms of political engagement.

Religion provides an important means of establishing personal identity and meaning, but fails, in the public sphere, to contest the power of the state. In the temporal arena, religion becomes a pawn of those who command armies, police, and prisons, and distribute wealth and honour.

Knowledge as power

Knowledge may be defined in both concrete and abstract ways. In its most concrete form, knowledge literally confers power. If two parties are on opposite sides of a legal dispute or otherwise engaged in direct competition, one may gain a power advantage through exclusive access to evidence, or may equalize power by obtaining information held by the opposing party. Nations in opposing positions devote much effort to acquiring accurate knowledge of the resources, tactics, and specific goals of one another, and use this knowledge to augment their power in case of war or aggression. In its more abstract forms, knowledge refers to mastery of or authority over a particular area of specialized information. Knowledge constitutes a position of power for the individual who has attained it. Those who have completed degrees in formal educational systems generally gain power in proportion to their knowledge, not only in terms of employment opportunities but also in self-realization and social status.

Copenhagen by Michael Frayn, which premiered at the National Theatre in 1998, places knowledge in the context of the race to develop powerful weapons in a war between nations. It speculates about the knowledge sought, shared, or denied in a meeting between two world-renowned physicists during the Second World War. Niels Bohr was a Danish theoretical physicist whose work contributed to the understanding of nuclear fission. Werner Heisenberg, a German physicist mentored by Bohr, authored the uncertainty principle, a foundational theory of quantum physics. The two men worked together in the 1920s, forming a close collegial relationship that acquired familial overtones because of Heisenberg's youth and Bohr's loss of his eldest son in a sailing accident. When Heisenberg returned to Germany he enjoyed high prestige and positions of leadership. During the war he was appointed director of the Nazi programme to develop atomic weapons. In 1941 Heisenberg visited Bohr, who lived in Nazi-occupied Copenhagen, constantly under threat because of his part-Jewish heritage. The play deals with the mystery of what questions were asked and what, if any, answers were given during this 1941 meeting.

On a bare stage lit to create an abstract image of the planet earth, and broken up only by three chairs, the three long-dead participants – Bohr, his wife Margrethe, and Heisenberg – discuss their memories of the meeting and related events. They all remember an encounter in which cordiality was strained by their opposing loyalties, and a walk proposed by Heisenberg, from which Bohr returned very upset. According to Bohr's recollection of the walk, Heisenberg attempted to pick his brains about nuclear fission and asked about Allied efforts to develop an atomic bomb. He says that Heisenberg queried him about his moral perspective on practical applications of nuclear energy and seemed to be interested in recruiting him to work on the German nuclear programme. Bohr relives his outrage at the idea that Heisenberg would consider him a potential collaborator with the Nazi regime.

Heisenberg's quite different memories of the meeting focus on what he describes as his effort to stall Germany's development of atomic weapons. According to Heisenberg, he sought assurance from Bohr that the Allies were not close to producing an atomic bomb so that he could continue stalling without worrying about endangering his country. Reminding his hearers that overt resistance to Hitler's regime was impossible, Heisenberg insists that the only way to oppose Nazi aims was through sabotaging them while appearing to give support. He further argues that he considered the understanding of atomic science at that time too indeterminate to be employed as military power. The two men also discuss their lives after 1941, of which objective records exist. Bohr escaped to the United States and worked at Los Alamos on the atomic bomb. Heisenberg was captured by Allied forces and taken to England, where he and other German scientists were detained and questioned. He eventually returned to civilian life, but became an international pariah because of his involvement with the Nazis.

Answers to the questions introduced by the play remain uncertain at its end. The two men offer contradicting opinions about what the other contributed to the course of the war, and neither can produce independent support for his version. Bohr refers to Heisenberg's ambition and youthful rise to preeminence in suggesting that he was compelled by a competitive drive to work feverishly on a project that, had it reached completion, would have melted down and killed him. Heisenberg feels bitter about the reproach he has faced over his role during the war. As he sees it, those who built the bombs that were dropped on Japan cannot claim moral superiority, and their refusal to shake his hand demonstrates the arrogance of the victor rather than a clear conscience. The play leaves unanswered the question posed to Bohr by Heisenberg in the contested meeting: what is the moral responsibility of someone with specialized knowledge that may be used to destroy? Both men confronted that question, but one was a closely watched insider, while the other was an outsider threatened by an oppressive, murderous, regime striving for world dominance. This difference of positionality implies that there is no way to answer such a question objectively, because there is no way of knowing what course each man might have pursued if he had been free to make any choice. The characters' interactions demonstrate that the human motives underlying events elude certainty. Rather than establishing facts or solving the mystery, the play leaves the audience with the conundrum of two different versions of knowledge: knowledge as too indeterminate to be employed as power, and knowledge that can be applied to mass destruction.

The History Boys by Alan Bennett, a 2004 production by the National Theatre, focuses on a secondary school classroom and therefore deals with a more generally accessible type of knowledge that manifests its power in individual lives. A celebration of teaching and learning, *The History Boys* uses Brechtian-style episodic scenes separated with songs and video sequences to explore the empowering effect of education. A group of sixth form boys, who exemplify the energies, aspirations, confusions, desires, and rivalries of male adolescence, are preparing for scholarship examinations in history at Oxford and Cambridge. To achieve admission to these institutions they must overcome the disadvantage of attending an unknown school in a northern city. The National's production effectively used a set with moving walls and scrim, as well as videotaped scenes, to suggest the hallways, classrooms, periods of intense activity, and intervals quiet typical a secondary school.

The two men who work most closely with the boys demonstrate contrasting, and to an extent conflicting, styles of teaching. The older teacher Hector has established his classroom as an independent realm devoted to learning for its own sake. He encourages the boys to learn poetry 'by heart' (48) and orients his teaching of language and literature towards enriching the students' personal understanding rather than equipping them for success. He considers examinations 'the enemy of education' (*ibid.*) and crosses educational boundaries, using class time for singing, dramatic improvisations in French, and a trivia game centred on old films. He also uses his power to cross personal boundaries, addressing the students with mock insults, hitting them with a lesson book in exaggerated shows of exasperation, giving some of them lifts on his motorcycle, and one-handedly fondling his passenger's genitals during the motorcycle rides. Hector's intense energy, his personal approach, and his passion for knowledge and understanding drive his love of teaching and inspire enthusiastic learning among his students. The boys eagerly look forward to Hector's lessons, even though they do not appreciate his motorcycle intimacies.

Disapproving of Hector's independence and determined to bolster his school's reputation through his students' exam results, the headmaster hires a new teacher to supervise exam preparation. He demands that Hector share his classroom with Irwin, whose approach to education proves antithetical to Hector's. Irwin considers knowledge a commodity to be acquired and applied to career goals, rather than a force that brings about change in individual lives. His reference to 'gobbets' of information that might favourably 'tip the balance' (48) in exams implies the manipulation of acquired resources. Irwin teaches the students to distance themselves from the emotional and moral implications of what they have learned in order to make attention-getting arguments that sidestep questions of truth. Thus, Posner, a Jewish student who originally reacts to a question about the Holocaust by speaking 'from the heart' (72) about the Nazis, later reports being praised for avoiding a tone of moral outrage in an essay about the Holocaust. Though strongly attracted to one of the students, Irwin does not act on his impulses. He does, however, mislead the students with his claim to have attended Oxford – a claim investigated by one of the students and found to be entirely false.

Hector's flawed character propels the two teachers to very different fates. The headmaster discovers Hector's inappropriate behaviour and demands his resignation. As the headmaster confides to one teacher, he is delighted at this easy means of dismissing a man whose teaching gets only 'unpredictable and unquantifiable' (67) results that cannot be used to bolster the school's reputation. Before his resignation takes effect, however, Hector dies in a motorcycle crash. The play ends with a series of moving tributes to Hector by the students. In an ironic twist, Irwin was Hector's passenger on the day of the crash; he survives the accident, though confined to a wheelchair. Irwin goes on from his teaching experience to become a media journalist and, later, a politician. He finds his wheelchair an advantage, observing that 'disability brings with it an assumption of sincerity' (60). Years later, when he is approached by Posner, he seems to have still found no passion or connection outside himself.

The boys, up to a point, exemplify a common fate. Throughout the play's scenes they display the self-possession, witty argument, and broad familiarity with English history and literature in which they have been schooled. Every one of the eight boys (improbably, but consistent with Bennett's own reported experience as a student at Leeds Modern School in 1951) gains a place at either Oxford or Cambridge. Moving on to careers in law, business, and teaching, they show the understanding and confidence developed in secondary school in various ways. The two students at opposite ends of the social hierarchy within the class – Dakin, the leader, and Posner, the misfit – indicate the range of outcomes that may result from the interaction of different educational approaches with the potential of particular individuals. The final scene reveals that Dakin, the student who most admired and emulated Irwin, becomes 'a tax lawyer telling highly paid fibs' (107). Posner, the only one of Hector's students who 'took everything to heart, remembers everything he was ever taught' (108), dropped out of university, lives alone, and suffers periodic bouts of depression. The achievements, friends, and memories of his secondary school years have remained at the centre of Posner's life. Despite their differences, one student seems to speak for the group when he says of Hector's teaching, 'Love apart, it is the only education worth having' (109).

Bennett uses the experience of education and the characters Hector and Irwin to express current conflicts about knowledge as power. Hector's desire to share heartfelt understanding and involve the students in experiences of truth that transcend time and situation references a philosophical approach that has become devalued in contemporary thought. The cynical and clever but formulaic manipulation of facts taught by Irwin, along with the emphasis on skills rather than knowledge and the primacy of personal ambition, may be associated with the effect of Thatcher's politics on British society. In actuality, these attitudes pervade contemporary life. The real question posed by *The History Boys* is not just how to learn, but how to live. It creates a contest between a humanistic framework for living, with its dimensions of seeking personal meaning and the common good, and an opportunistic framework, with materialistic values and no ethical commitment. It furthermore reintroduces history as a process through which to create meaning. Its generational divide reminds audiences that each generation brings to its education multiple potentials for individual advancement and genuine discovery. Its teachers need not be flawless to inspire their students' learning and support their development of maturity and understanding. The students need not be exceptional individuals or members of the social elite to value learning and incorporate it into their lives rather than merely using it to attain goals.

The plays that consider power in abstract forms employ enquiry and analysis rather than argumentation. They eschew advocacy, approaching their subjects with questions and scepticism. While dealing with political issues, they avoid topicality in favour of looking at historical developments. They call attention to the operation of different kinds of power within society and also to the limits of these forms of power. They explore mechanisms of leadership and control, patterns of personal ambition and loyalty, and the forms and institutions through which humans search for truth and establish meaning. The dominant concern in this group of plays is not the acquisition of power or the rebellion against authority, but rather the search for a context beyond the self in which to connect one's life and its endeavours to a sense of significant and enduring meaning. The search for such a context implies the desire for political thought grounded in broad understanding of humans and their societies rather than narrowly instrumental in manipulating policies and problems. This desire constitutes a notable but as yet undeveloped strain of idealism in post-Thatcher political theatre.

5
Issues for Post-Thatcher Britain

The re-emergence of issue plays in Britain since 1998, and acceleration of the trend from the year 2000, unequivocally signals theatre's return to engagement with politics. This fresh engagement may be seen as the bursting of a dam after years of mounting dissent and frustration. Alternatively, the renewed engagement with politics may be seen as the outcome of several years of more accessible and responsive government, along with increased confidence in the possibility of political action on issues of urgent concern. Plays about specific political issues arise from a strongly felt need for change that impels proponents of such change to attempt to defy the odds against success as they bring their case before the public. Anger and even desperation propels such plays. However, an issue-oriented play also relies on some level of confidence that the political system will attend to the message and has the capacity to change. The systematic and thorough control exercised during the Thatcher years denied a hearing to issues of concern to the opposition. Under these conditions, anger turned to frustration, despair, and disengagement from public affairs. Recovery from these conditions did not occur overnight. Neither the in-yer-face plays nor the responses to them articulated specific issues that could conceivably be addressed by the government. The Labour government in power since 1997, however, seems to have restored the level of confidence necessary to present arguments for social and political change, even though a number of the issue plays call for changes in that government.

Recent issue plays begin with a specific political issue but go beyond its specifics to explore its implications for the direction of national life. Some of the issue plays develop into discussions of the state of the nation, identifying disturbing trends in the conduct of politics and erosion in key national institutions. The issues that surfaced first were the critical ones of racism in British society and the situation in Northern Ireland. As the Blair government has made controversial decisions, its dynamics and style have given rise to a number of plays. Issue plays have been brought to the stage by both new and established playwrights. Notably, playwrights have turned to documentary modes of presentation, allowing events themselves to speak to the audience in a direct and immediate style.

Racism

The issue play returned to Britain with *The Colour of Justice*, by Richard Norton-Taylor, performed at the Tricycle Theatre in 1998. Norton-Taylor, a journalist with the *Guardian* since 1975, who considers himself as editor rather than playwright, structured the play by selecting passages from the transcripts of the Stephen Lawrence Inquiry. This inquiry generated 11,000 pages of testimony during a period of 69 days in 1997, and culminated in the issuance of the Kent Report in December of that year. It concerned the fatal stabbing of a 17-year-old black youth, Stephen Lawrence, who was assaulted at a bus stop in southeast London on the evening of 22 April 1993. A friend who was with Lawrence but outran the assailants said that a gang of white youths shouted a racist epithet and attacked without provocation. Though five young white men in the area who were known for racist behaviour and unlawful activity were arrested and eventually charged, the charges brought against these suspects were dropped because of what was considered weak evidence.

The case captured public attention and exposed racial tensions in London, setting in motion a large anti-racist demonstration, a march in south London by the white supremacist British National Party, and an appearance by Nelson Mandela. Lawrence's parents were brutally harassed, and they buried him in Jamaica out of fear that a London gravesite would be vandalized. Frustrated by the failure of the government to prosecute, the parents instituted private proceedings in 1995. They were unable to secure convictions, because, as before, the eyewitness testimony of Lawrence's friend, Duwayne Brooks, was discounted. The inquest was reopened, and the Coroner's Jury returned an extraordinarily specific verdict of unlawful killing 'in a completely unprovoked racist attack by five youths' (*Guardian* 14 February 1997; quoted in Norton-Taylor 11). In response to the inquest verdict, Britain's Home Secretary met with Lawrence's parents and promised a public inquiry into the failure to bring his killers to justice.

Norton-Taylor's dramatization, which aims to present 'as fair, balanced and as rounded a picture as possible' (6), targets the police investigation of the murder. Director Nicolas Kent, working with a cast of 27 actors, gave the play a deliberately undramatic tone and set it in an accurate representation of the room used by the inquiry. Questioning begins with the two on-duty police constables who responded to the first call for help and proceeds up through the law enforcement hierarchy to the Assistant Commissioner of the Metropolitan Police. The questions uncover mistakes and missed opportunities at every level of the police investigation, including failure to properly search the area, follow up on testimony by eyewitnesses, and monitor the activities of the prime suspects. The inquiry also establishes the absence of written records and strongly suggests the possibility that notes were destroyed in an attempt to cover up police incompetence or even collusion with the murderers. Police officers and officials repeatedly deny that their actions were influenced by racist attitudes, but the denials ring hollow in the context of an entirely white police force, the black parents who were not treated with customary sensitivity, and the black friend whose identification of the suspects was not considered strong enough for conviction. The inquiry reveals failure by the police to protect potential witnesses from intimidation by the suspects. It also uncovers a possible cause for the inadequacy of the investigation and the missing records: a police officer assigned to the area had been associating with a known criminal who was the father of one of the suspects. It does not present indictments for the murder of Stephen Lawrence, although it goes over the evidence that has consistently pointed to the five initial suspects. The silence with which the play ends denies the closure expected of both murder cases and dramas.

Tribunal theatre uses words from the documents of an official proceeding, but its primary function as theatre is to produce the 'restored behaviour' conceptualized by Richard Schechner as the central aspect of performance (1985: 35). In this play, the performance of an action often produces the most telling evidence of police attitudes. As witnesses shift uncomfortably, avoid meeting the gaze of their questioner, or pause uncertainly before speaking, they create an unspoken narrative that may illuminate, run counter to, or provide context to their words. Witnesses may offer telling contrasts, as is the case when a neighbourhood couple who happened on the scene describe kneeling by the victim to render aid and comfort, while the police officer who responded to the emergency did not touch him to determine his condition. A particularly dramatic moment occurs when the chief counsel to the inquiry questions William

Ilsley, the superintendent overseeing the investigation, about a written list of suspects handed to him by the victim's mother approximately one month after the murder. Mrs Lawrence had been distressed by Illsley's action of crumpling the note in his hand as though to throw it away. Ilsley denies having crumpled the note, referring to his action as folding. At that point, the counsel brings out the note, with all its creases, and asks Ilsley to reproduce his action. He does so, folding it repeatedly until it is a tiny morsel of paper in his hand, unmistakably communicating an intention to disregard the information.

The bereaved parents, Doreen and Neville Lawrence, are the tragic figures in this drama. The play conveys their grief over the loss of their intelligent, hardworking, and law-abiding son, and their anger that his murder was not treated with seriousness and urgency. The inquiry justifies their outrage by making public the attitudes of police officers at the scene who assumed that the injured teenager had been fighting and paid scant attention to the nature or extent of his injuries. By systematically compiling evidence of the incompetence and lack of accountability among the police, the inquiry builds a case that prompts, at the very least, questions about the commitment and ability of the police force to protect the lives and respect the rights of black citizens.[1] The inquiry cannot provide the justice they have been denied, but it documents the injustice and provides a belated but necessary demonstration of public respect for the Lawrence family, in the attention given to their written statements, the apologies offered by the authorities, and the periods of silence dedicated to them at the beginning and end of the inquiry.

The Colour of Justice, though not the first example of tribunal theatre,[2] made a major public impact. It stands out as one of the most successful attempts to use theatre to make an issue visible. After its widely praised run at the Tricycle, it played for an additional two weeks at the Theatre Royal Stratford East, then transferred to the Victoria Palace in the West End. In 1999, the play toured nationally, concluding with a one-week run at the National Theatre, which featured discussions with key figures in the Lawrence case after every performance. The play was made into a BBC film in 1999. It extended the effect of the inquiry, fuelling widespread dissatisfaction with the police investigation of the Stephen Lawrence murder, exposing institutional racism in the police force and other public agencies, and increasing pressure for meaningful changes. A law passed in the wake of the inquiry, the Race Relations Act of 2000, requires public agencies to demonstrate non-discrimination in their employment practices. Most recently, the Independent Police

Complaints Commission has announced a new inquiry following the release of a BBC documentary in July 2006 that made specific allegations of corruption against an officer central to the investigation, through a witness who claims that the officer was paid to suppress evidence.

In addition to influencing public opinion regarding the Stephen Lawrence murder case and the police as an institution, *The Colour of Justice* brought attention to a new form of political theatre. Tribunal theatre brings issues directly to the public, giving the audience an immediate experience of the conflict and inviting their individual judgement. Eschewing the elements of entertainment long considered to be indispensable to successful theatre, it condenses and translates lengthy and sometimes densely worded official proceedings into a form that can be viewed and comprehended in less than three hours. The tribunal form relies on existing structures of feeling (see Williams 1961: 48) that privilege factual presentations in regard to political issues. These plays bypass both government and mass media, which in some cases have become discredited, to bring issues before the court of public opinion. *Independent* theatre critic Paul Taylor sees tribunal plays as the result of an intrinsic tendency in theatre to become a courtroom, sug- gesting that theatre, 'by its very form, offers a heightened image (often painfully lacking in the real world) of people being held to account in person and in public' (10 June 2004). The appeal of tribunal theatre lies in the opportunity it gives the audience to consider individual actions, see and hear those individuals defend their actions, and participate in judgement. This means of presenting issues thus functions in a unique way to involve audience members. The evidence may be vivid and even dramatic, but the memorable experience consists in drawing one's own conclusions on the basis of hearing and considering the evidence.

Norton-Taylor, who has been the primary editor of tribunal plays, has written about the way in which they differ from standard journalism.[3] In his view, the value of the play lies in the directness of using only words spoken in the proceedings, rather than second-hand descriptions, and the completeness of a finished story, as opposed to the day-to-day reporting of pieces of the story. For that reason, he finds it necessary as a playwright/editor to pay more attention to the overall shape of the piece rather than include headline-generating moments. He has found subtleties and nuances particularly revealing in hearings that involve lengthy verbal exchanges. Norton-Taylor stresses the importance of not over-dramatizing, and especially not caricaturing the characters, and mentions that some actors cast in the 2003 play about the Hutton Inquiry attended the hearings. He considers the theatre 'a medium complementary to newspapers' that 'can lead to a greater understanding of how we are governed and what is being said and done on our behalf' ('How Hutton Became a Play').

Events such as the Brixton riots of 1981, 1985, and 1995 and the Bradford riot of 2001, as well as the Stephen Lawrence murder and the unsolved 2001 murder of ten-year-old Damilola Taylor on a south London estate have prompted increased awareness of racial divisions and threats. In *Sing Yer Hear Out for the Lads*, Roy Williams shifts his focus from issues of identity within the Afro-Caribbean community to exclusion of Afro-Caribbeans from the concept of British identity. Originally produced in 2002 as part of the Transformation Season at the National Theatre, the play was remounted by the National in 2003; after touring it returned to the National in 2004. The action of this realistic play occurs in the context of an actual sporting event, the World Cup qualifying match between England and Germany in 2000. Regulars gather in the King George pub in southwest London, and a large television screen presents the game, which England lost 0–1. The group assembled around the screen includes both black and white young men ostensibly united in pride and support for England. It includes several members of the pub's football team who won their match earlier in the day mainly through scoring by the black player Barry. The men watching the match cheer together and confront one another, while the bartender Gina and her friend Lee, an off-duty policeman, attempt to keep the situation under control.

Using football as a touchstone of identity and action highlights the illusionary aspect of national unity and the continual threat that identity may be expressed through violence. From the beginning of the play, remarks and actions reveal that the white characters define themselves in terms of difference from blacks and quickly turn to racial stereotyping if a person of colour annoys or angers them. Two of the white men display explicit and outspoken racism. Lee's unemployed brother Lawrie vents his inarticulate resentment in crude insults and threats, while the educated and deliberate Alan voices white supremacist propaganda and ominously intones 'rivers of blood' (58). Others display casual racism, calling Barry, the pub team's black player, 'boy' and rubbing his head. Gina tries to suppress their racist remarks in the presence of her black customers; however, when her teenage son arrives, having been robbed of his mobile and jacket by a black youth, she also resorts to racist stereotyping. Another black acquaintance returns the stolen items, but is roughed up anyway by Lawrie. The two black men in the pub crowd show contrasting attitudes towards the white group. Barry, with the flag of

St. George painted on his face, just wants to be one of the gang, ignoring the disrespect from his white companions while attempting to bond with them through conspicuous cheering. Mark, watchful and serious, stands apart from the group, ready to protect his younger brother Barry and hoping to persuade him to leave the pub.

Tension escalates when Sharon, the mother of the youth who returned the stolen items, arrives at the pub to complain of the way her son was treated. Lee tries to subdue her when she becomes belligerent. He calls on Mark to help him, but Mark refuses to get involved, and the pub owner rings the police. When the police arrive, Mark tries to help Sharon. Amid this conflict, Barry concentrates on the game, cheering more loudly to drown out distractions. Alan deliberately provokes arguments with Barry and Mark, maliciously pretending sympathy in order to deliver his racist polemic. Lawrie appears eager to fight, while his brother Lee keeps reasoning with him. When England loses and the mood turns foul, Lee sees that he will not be able to control his brother, and warns Mark and Barry to leave. They refuse, but just then a brick comes through the window, thrown by a group of black youths. Lee goes out to confront them, and this time Mark joins him. They temporarily forestall further violence, but now cannot leave the pub.

The play climaxes with Mark being stabbed to death in the toilet. Despite Lawrie's many threats, he does not commit the murder. Glen, Gina's frightened teenage son, who has armed himself with a knife and resists Mark's attempt to take it from him, kills Mark. In the confusion that ensues, when Mark's bleeding body is found, Lee accuses Lawrie and begins an arrest procedure, but then Gina's son is discovered with the knife. Lee turns his attention to Barry and, despite Lawrie's jeers, attempts to calm and console him. Barry and Glen, in the final moments, pour hatred on each other's racial group. As they scream at each other, Lee remains alone in the middle of their conflict, and a riot begins outside.

The play's setting and situation define its characters in a way that limits their range of identities. All are football fans, and all support England against Germany. All are similarly young, their age giving an ironic dimension to the taunt shouted at the screen, 'Stand up if you won the war' (78). Their support for different teams within England, as well as their differing attitudes towards particular players, suggests racial and regional loyalties. Different kinds of temperament are evident, with some men prone to violence while others offer no threats, and some talking expressively while others are mostly silent. Some men are employed, some not. Some play on the pub team; others do not. Some focus closely on the televised match, while others seem more interested in social interaction. Some bring a family identity to the pub, because they are there with another family member. Only Lee, the police officer, brings his work identity to the situation. In part because of this professional identity, only Lee holds on to his neutrality when all other identities collapse into racial division. This collapse dooms Lee's hope of maintaining his friendship with Mark, Barry's hope of being accepted by the other pub regulars, and Glen's desire to be accepted by black peers.

Racist hatred appears like a raging fire, ignited by accidental and deliberately produced sparks. The crucial spark comes not from the belligerent Lawrie or the sly Alan, but from an insecure adolescent yearning for acceptance by the black youth 'Bad T' and his friends. Once racist violence breaks out, it traps all the men – even those who have said almost nothing all evening – in the pub, and metaphorically in the structure of racism. At the end, all nuances of individual identity dissolve in the division between white and black. The play offers little hope for progress in race relations. Those who are moderate, peaceful, and hopeful of friendship across racial lines become victims of those who through a deliberate effort to hold on to their privileged status or through an atavistic form of loyalty define themselves in opposition to others.

Gladiator Games by Tanika Gupta, which premiered at the Crucible Studio in 2005, deals, like *The Colour of Justice*, with the killing of a young man in the context of institutional racism. In this case, Zahid Mubarek, a young Asian man due to be released after serving a short sentence for minor offences, was killed by his cellmate while in custody at Feltham, a prison for young offenders. The cellmate was known to be severely disturbed and violently racist. This play not only uses verbatim testimony from an inquiry into the death and from interviews with members of the family, but also includes fictionalized dramatic episodes, based on letters and family memories, of Mubarek in prison interacting with his cellmate, a friend, and his visitors. It creates a sympathetic portrait of a young man who had made mistakes, learned from them, and was hoping to get a job and restore his parents' pride in him when he got out of prison. In considering the injustice of the young man's death, the play goes beyond accusing the prison authorities of negligence to make the shocking suggestion that Mubarek was deliberately placed with a violent racist in order to provoke a fight. An informant tells of hearing about 'a game played by prison officers in Feltham called Gladiator or Coliseum', in which they would place together prisoners who were expected to attack one another and bet on the outcome (72). Authorities dispute this allegation, and the person who made it subsequently admitted that he exaggerated and spoke without evidence. In the end, the death was blamed on insufficient staff in the prison, but family members could not forget the more sinister accusation.

While this play sensationalizes the death of Mubarek with an unsubstantiated allegation and fictionalizes his character in the dramatic episodes, it nevertheless points to serious questions about the value accorded to the lives of Asians in Britain. The young man who died committed no violent offences and showed promise of reform. His release was to occur the very day he was murdered. While he was in custody he had no means of defending himself, and his family had no means of protecting him. His life was literally in the control of the authorities, who failed to preserve it. Though this failure was not shown to be deliberate, it demonstrated serious negligence, since the cellmate had established a record of violence and potential to do harm. Such negligence points to a system that does not value the lives of prisoners, especially those of colour. The play includes a lengthy statement by Colin Moses, Chair of the Prison Officers' Association. Moses, a black man himself, says he 'felt touched by that death' (53) and made a personal apology to the family. Moses details the disproportionate number of non-white men in prisons, compared to their presence in the population, the difficult conditions of prison, and the lack of black and Asian prison staff. He calls for changes, not just in the prison system but in society as a whole, to change a social system in which 'we send more young black men to prison than we do to university' (54).

These issue plays demand attention to the problem of racism in British society. Their common emphasis on the murders of young men urgently points to the threats faced by non-white citizens, especially young Afro-Caribbean and Asian males, in pursuing the ordinary activities of daily life. Highlighting cases in which government officers have failed to defend the lives and rights of black people, they call for recognition of the racist attitudes infecting institutions charged with protection of the public. These plays make an implicit but strong appeal for action, and their appeal has met with some success. The anti-discrimination law enacted in 2000 indicated public responsiveness to the problem and willingness to take action to end institutional racism in governmental organizations. At the same time, the lack of closure in the plays, as well as disturbing events, such as the Damilola Taylor murder, suggests the need for further changes to move Britain towards equality and justice for all its citizens.

Northern Ireland

The conflict in Northern Ireland remains one of the most intractable issues within Great Britain. Since the Good Friday Agreement of 1998, the peace process has dealt primarily with the issues of Irish Republican Army (IRA) disarmament and government legitimacy. These issues highlight the difficulty of power sharing and reconciliation in the face of decades of opposition, suspicion, and violence. Gary Mitchell has offered a significant series of dramatic commentaries on these issues. Born in 1965, Mitchell grew up and still lives in the Protestant, Loyalist, working-class community of North Belfast, gaining perspective on Northern Ireland's recent history at first hand. He focuses on internal pressures and divisions within his community that have created new concerns in the context of changes since 1998. Mitchell's plays differ from other recent issue plays in their use of fictional drama and their assumption of a shared frame of reference – Loyalist philosophy, history, and current position in Northern Ireland's ongoing conflict.

The Force of Change, first performed at the Royal Court's Theatre Upstairs in 2000 and then transferred to the larger downstairs theatre in the same year, explores recent changes in the Royal Ulster Constabulary (RUC), the police force charged with confronting violations of the peace. Changes in the force have brought in women, such as Caroline Patterson, the detective sergeant up for promotion who leads the unit presented in what might be a typical day of work. Her three male colleagues display different attitudes towards working with a woman, from Mark's helpful support to Dave's seeming neutrality, to Bill's undisguised hostility. In action over the course of a working day, the four interrogate two suspects. The first, most important to the interrogation team, is suspected of political violence. This man, Stanley, has been involved with the Ulster Defence Association (UDA), a Loyalist paramilitary organization, and has been accused of intimidation aimed at preserving the power of the UDA within the Loyalist community. He was arrested after eyewitnesses reported that he shot one man in the knees and extorted money from another at gunpoint, but the witnesses have since withdrawn their accusations. He refuses to answer questions, and will be released at the end of the day if the police cannot get enough evidence to charge him. The second suspect, a teenager who has been caught after a high-speed chase with a stolen car, seems to have no connection with politics. Nevertheless, the police pressure him to offer evidence against the UDA man.

In this contest between the RUC and the UDA, the UDA wins easily. At one point, Stanley is found to have a mobile phone belonging to Bill. Though Bill admits collusion, Caroline cannot dismiss or lodge a complaint against him, because it would appear to be retaliation and might interfere with her promotion. The suspect has used the phone to identify

Caroline's personal vehicle and home address to his associates, thereby placing her family in danger. At the end of the day, the UDA man walks free, while Caroline's family has been sent into hiding. In utter frustration, David, the youngest of the officers, can only try to make freedom uncomfortable for Stanley, threatening to grab him some night as he makes his way home from his 'favourite loyalist club' and drag him into 'the nearest dark alley' (80–81). Mark had earlier threatened the teenaged car thief with the vigilante justice of the street in an attempt to get information from him, and finally, in frustrated rage, attacked him physically.

These encounters reveal a police force unable to enforce the law in a society where an illegal organization is more powerful than the legal government. Their frustration at being unable to act effectively leads the officers to resort to violence and intimidation. In the end, the real victory for lawlessness consists in the fact that the police officers adopt illegal methods. The only one who remains within the boundaries of law is Caroline, but as a woman in a male-dominated organization, and with her family forced into hiding, she appears extremely vulnerable. In a larger sense, the play shows the corrosive effects of the conflict in Northern Ireland. People trapped by this unending division find their world limited, their choices defined, and their views warped by the conflict that permeates their community and every aspect of their lives. Though written from within the Loyalist community, *The Force of Change* appeals to an audience outside it to recognize the need for re-establishing the rule of law and effective structures for its enforcement in Northern Ireland.

Loyal Women, produced by the Royal Court in 2003, highlights the role of women in the activities of the Loyalist paramilitary, UDA. Its characters include four generations of women in one household. At the head of the household is the steely, hard-working, and endlessly self-sacrificing Brenda. At the age of 35, Brenda is responsible for her teenaged daughter, the daughter's infant, and a bedridden mother-in-law. The half-senile mother-in-law, lazy daughter, and demanding infant make Brenda's life difficult enough, but early in the action the local leader of the women's UDA pays an uninvited visit with the news that Brenda has been nominated to take over leadership of the unit. Brenda does not want this position, but has little choice. Her life has been directed by the UDA since she was a teenager and, at their direction, killed a woman suspected of association with the IRA. Now she becomes involved, despite her wishes, in the nightmarish punishment of a young woman who has brought her Catholic boyfriend into the Protestant area. Brenda tries to prevent the violence threatened by her two rivals in the organization, and simultaneously tries to discourage her own daughter Jenny from becoming involved with the UDA. Brenda's husband, who served a long jail term after admitting to the murder she committed, has come back and wants to move in again; meanwhile, her friend Mark would like to have a sexual relationship. Brenda, however, excludes both men from her already overburdened life.

The play shows Brenda attempting to maintain control of her life and protect her daughter and granddaughter. She is a strong woman who does not waver in her dismissal of the husband she once loved but now associates only with sexual betrayal. She also refuses to marry Mark, remaining within the bounds of a female family and community. Although she has changed since her 'head-banger' stage as a young woman, learning 'to think for myself and look after myself and how to prioritise' (85), she cannot banish the UDA from her life. The organization has decided that Brenda's moderate views and social skills provide the 'friendly, sociable, politically-correct face' (79) it needs in the present situation. Though reluctantly accepting the responsibility of treasurer, she strongly resists appointment to the chair position, offering to resign in favour of a rival. Complications within her family, however, combine with the UDA's power to keep her involved. She fails to prevent violence or keep her daughter out of the action. The open ending leaves it uncertain whether Brenda will ever be able to enjoy peace.

This play joins with previous work by Mitchell, especially *Trust* (1999), to emphasize the personal cost of the political division in Northern Ireland. The most recent period of stability and movement towards political negotiation has brought changes and the need for adjustment on both personal and societal levels. Among the changes is attention to the domestic sphere and to the lives of women. Women like Brenda and her daughter may now consider their own needs. Nevertheless, as the play demonstrates, both family responsibilities and the pressures involved in being part of an insular community with unresolved grievances make it difficult to find physical and conceptual space in which to create an individual identity and function autonomously. While Brenda has moved towards that conceptual space, there is no certainty that she, her daughter, or granddaughter will be able to live in it.

The most recent play addressing the conflict in Northern Ireland is *Bloody Sunday: Scenes from the Saville Inquiry*, edited by Richard Norton-Taylor and first performed at the Tricycle Theatre in 2005. A tribunal play based on the official inquiry carried out in Londonderry and London from March 2000 to November 2004, *Bloody Sunday* deals with an event widely considered to have played a major part in the most recent chapter of Catholic–Protestant opposition in Northern Ireland. The Saville Inquiry was the second official examination of the 1972 incident, in which 26 civil rights marchers were shot – 14 fatally – by British army paratroopers. It was instituted because the first, the Widgery Inquiry, conducted in the immediate aftermath of the event, reached conclusions that were widely disputed.[4] A previous play about the Bloody Sunday killings, Brian Friel's *Freedom of the City* (1973), expressed the belief that those hearings were a cover-up by using their statements as an ironic counterpoint to its dramatization of the last hours of three individuals killed that day. Families of the deceased had long campaigned to re-open the inquiry, and their demands gained credence when documents released by the National Archive indicated bias on the part of the presiding official and failure to examine discrepancies in the testimony of soldiers involved.[5] As counsel to the inquiry, Christopher Clarke, QC, stated the purpose of the proceedings: to recover the truth, which had been 'the first casualty of hostility' (7). Establishing an unbiased and factual record of what occurred, even after a time lapse of nearly 30 years, was understood to be an important step in reconciliation.

The scenes, which present 12 witnesses, seek answers to the most urgent questions left unanswered by the previous inquiry: were any of the marchers firing at the army, and did some of the soldiers act irresponsibly or illegally? Among those who became caught up in the march was Bishop Edward Daly, who testifies that as he made his way home from mass that Sunday, he observed youths throwing stones at the soldiers, heard shots nearby, and began running with the panicked crowd when paratroopers in armoured vehicles began driving towards the marchers. Daly was very near an unarmed youth who was shot from behind and killed, and he remained to aid and administer last rites. He states that there was 'no threat posed to the army at the time they opened fire' (13), and further testifies that he saw another unarmed man shot, while a man firing a small handgun at the soldiers escaped their notice. A marshal for the march tells of trying to control the stone-throwing youths, and then of being shot when he remonstrated with soldiers over killing an unarmed man. The well-known former MP Bernadette Devlin McAliskey testifies to the non-violent aims of the march. She describes paratroopers firing on the assembled crowd of 30,000 just as she began her speech. Questioned about the potential for violence among the IRA factions, McAliskey asserts that the government 'ordered the Army to shoot the citizens' (30) and should be brought to the International Court. Additional eyewitnesses describe soldiers firing indiscriminately on the marchers, responding to

stone-throwers with deadly bullets, shooting a man waving a white handkerchief, throwing the wounded and dead into an armoured vehicle, and
preventing first-aid workers from rendering aid.

Testimony from those involved in the march, which consistently
claims that the demonstrators carried no guns and did not provoke the
shooting, often draws applause from observers. The testimony of representatives of the army, from a general to anonymous soldiers,[6] often
provokes shocked silence. The general in charge of Land Forces in
Northern Ireland describes hatred and fear between the Catholic community and the army. He admits to considering a policy 'to shoot
selected ringleaders among the DYH [Derry Young Hooligans] after
clear warnings' (52), but insists that the soldiers fired only after being
fired upon, and denies knowledge of a man killed while waving a white
handkerchief and attempting to aid an injured person. The paratroopers'
commander reveals a poorly conceived plan to arrest hooligans and
concedes that herding marchers with armoured cars made it impossible
to separate hooligans from non-violent participants or bystanders. The
officer who directed operations on the scene retracts a previous claim to
have seen 'a man with an M1 carbine on the balcony of a flat' (74).
Questioning reveals numerous inconsistencies in this officer's previous
statements about where he was and what he observed. The first anonymous soldier retracts statements made to the earlier inquiry that he saw
home-made bombs thrown at the soldiers and a gunman firing from a
window, explaining that he was 18 years old and frightened by the
military police when he made the statements. The second soldier admits
to killing four people, but even when confronted with photographic
evidence to the contrary insists that he shot only at people with guns
or bombs. The Official IRA commander who gives evidence also retracts
his earlier statement, admitting that some members of his unit had guns
and that shots were fired, but maintains that they fired only in response
to the army's shooting.

The play ends abruptly after the IRA commander's testimony, with no
summary statement or gesture of closure. What emerges through the
progression from tragic images and sweeping indictments to reluctantly
conceded errors, attempts to maintain indefensible statements, abject
apologies, and admissions of perjury defies a simple explanation. Clearly
the army did not uphold its responsibility to protect the innocent.
Unarmed and unresisting people were shot and killed. The army arrived
at the scene primed for conflict, expecting violence, and planning to
arrest hooligans. Confused about how to make arrests, they herded the
crowd down a street, using armoured vehicles, and then began firing.

With the retraction, revision, and demonstrated inaccuracy of the soldiers' testimony to the first inquiry, no evidence that the army was fired upon remains. Several witnesses independently describe the aggressive and confrontational style of the general in charge of the army in Northern Ireland. While evidence points to unjustifiable use of force by the army, it does not exonerate the IRA. There were, according to the IRA commander's belated admission, three armed IRA men among the crowd. Although the IRA had nothing to gain from disrupting a peaceful march that included their family members and demanded their civil rights, its guns contributed to the violence. Bloody Sunday's toll, as McAliskey notes, exceeds the 14 innocent citizens killed and the 13 wounded. The attack initiated three decades of violence and thousands of deaths.

The play was produced after hearings in 2004, but before findings had been published. A final report, which must consider oral statements by more than 900 witnesses and 2500 written statements, has not yet been issued (see www.bloody-sunday-inquiry.org/). The absence of a clear conclusion, coupled with the time that had elapsed since the original event, the length of the inquiry, and long-entrenched positions of partisans, muted the impact of *Bloody Sunday* in comparison with other tribunal plays. It received high praise from critics and ran to full houses, but did not tour. It was written about in the Irish press but has not been performed in Ireland. This play's short life on stage, as well as the inquiry's prolonged deliberation, may signal the desire of a weary public to put the Northern Ireland conflict out of mind. Nevertheless, it gives dramatic visibility to crucial issues in an unresolved political conflict. The tribunal play, as did the inquiry, reawakens anguish. It offers a compelling moral spectacle, as humble people stand up and speak in a public forum. It reveals shocking callousness and ineptitude in the army hierarchy. Rather than reinforcing entrenched positions, however, the scenes unsettle those positions, provoking thought about the various parties involved and the complex interactions among them.

These plays about the political situation of Northern Ireland urge new interpretations of an issue that has long been visible. They present unfamiliar perspectives and new information, implicitly calling for reassessment of the situation and fresh approaches. Mitchell's works reveal the fraying of long-standing alliances, as participants weary of isolation and struggle. The tribunal play, too, shows previously maintained lines crumbling, through the effects of time and distance, while also demonstrating that some participants in the conflict – especially those who maintain power by doing so – repeat the same provocations and justifications that they have offered for decades. While the conflicts between Catholics and Protestants and within particular factions have migrated to the back pages of newspapers, playwrights' engagement with these issues helps remind the public that they have not been resolved.

Political leadership

Recent examinations of political leadership have revolved around the governing style and decisions of Tony Blair, who became leader of the Labour Party in 1994, and in 1997 led the party to its first election victory in 18 years. Winning a second term as prime minister in 2001 and a third in 2005, Blair has served the longest continuing period of any Labour prime minister. Blair's government has pursued a centrist course, adopting the term New Labour to distinguish their policies from the Marxist-influenced philosophy of Labour's past and proclaiming the intent to find a 'third way' alternative to oppositions between left and right. Despite the landslide win in 1997 and personal popularity that began to decline only in response to Britain's involvement in the Iraq war, Blair has been criticized for his use of spin to manipulate public opinion, his closeness to an American president strongly disliked in Britain, and his refusal to overturn policies or reverse trends set in motion by previous Conservative governments. New plays have questioned the current political leadership in terms that employ satire, documentary evidence, serious dramatization, and tribunal form.

Feelgood by Alistair Beaton, a writer experienced in television but new to the stage, satirizes the Blair government's obsession with spin, the practice of interpreting policies and actions to coincide with popular opinion. It was produced by Out of Joint, first performed at the Hampstead Theatre in 2001 and transferred to the West End. Set in a hotel suite in Brighton or Blackpool during the annual Labour Party conference, the play shows fictionalized press secretary Eddie engaged in behind-the-scenes preparations for the prime minister's speech. Surrounded by the sleek paraphernalia of modern technology, which contrasts with the hotel's dowdy décor, Eddie and the speechwriter Paul painstakingly revise the speech. They receive, via the prime minister's personal assistant Asha, orders to include an upbeat reference to global warming and some jokes contributed by a television sitcom writer.

Outside the room a variety of issues and incidents threaten the unity of the conference. Demonstrators in the street carry on a noisy protest against capitalism. A group of delegates opposed to genetically modified foods takes over the microphone on the conference floor and questions the government's assurance that no secret trials of genetically modified crops are taking place. George, a junior cabinet minister, wanders into the suite with bad news: a company in which he has heavily invested has been conducting secret trials of wilt-resistant hops. By mistake, the experimental plants have been harvested and used in beer production. Males drinking this beer are developing female-type breasts. The journalist prepared to break the story is Eddie's ex-wife. As Eddie and George contemplate this development, Simon, the joke writer, arrives, adding a farcical element to the action with his ill-timed attempts at humour. Since Simon has overheard George's revelations, the increasingly frantic Eddie tries to keep him in the room without including him in discussions, repeatedly sending him to wait in the toilet, from which he emerges at inopportune moments.

The heart of Beaton's satire lies in its characterization of the various political operatives. Eddie, the power-obsessed press secretary modelled after Peter Mandelson, Blair's widely reviled architect of spin, dominates the action. Bullying and manipulating, Eddie exerts control over everything from the condition of the carpet on the speakers' platform to statements by delegates on the conference floor. Eddie will stop at nothing to maintain power. To keep Paul focused on the speech, Eddie tries to prevent his hearing about an accident that has resulted in his daughter's hospitalization. To deal with the beer fiasco, he blackmails George into taking the blame and resigning from the cabinet. By contrast, the young speechwriter Paul seeks to act on principle. He conscientiously refuses to help Eddie smear the characters of party members who spoke against genetically modified foods. Eddie, however, blackmails Paul by threatening to publicize his homosexuality. Some characters seek contact with power on any terms. Asha, the ambitious personal assistant who oversees the prime minister's wardrobe, revels in her closeness to him. George, the alcoholic party hack repeatedly hit by the protestors' improvised missiles, and Simon, the clueless outsider thrilled by proximity to the prime minister, allow Eddie to push them around without insisting on personal dignity.

Eddie's only real opponent is his ex-wife Liz, the journalist who threatens the party's future with her story of the experimental hops and feminized beer drinkers. Eddie opposes her without regard to legality, breaking into her online communications and her hotel room. Liz holds her own, demands a private meeting with the prime minister, and emerges from it with his promise to enact a full slate of traditional Labour measures, from raising taxes on the wealthy to repealing the Asylum Act, in exchange for her pledge to suppress the story. When

Paul reproaches her for making a deal, she argues that even the most praised piece of journalism seldom achieves more than the creation of a task force, while this deal will 'get rid of... a particularly nasty Act of Parliament' (94). Paul questions the eventual outcome, observing that the prime minister 'doesn't keep promises, but he doesn't break them either. He just erodes them. So you'll never be sure when you've been betrayed' (*ibid.*). He persuades her to take back the computer disk she had turned over to Eddie.

The party conference ends explosively as the demonstration outside becomes a riot, rioters invade the hotel, and tear gas canisters fly. When the smoke clears, the prime minister, who has throughout been referred to familiarly as DL – for divine light – by his subordinates, gives the speech. He opens by saying that the unexpected events have compelled him to throw away his prepared text and to speak 'directly, simply, spontaneously, from the heart' (103). The speech, with the slogans and weak jokes worked over by Eddie and Paul, is anything but spontaneous, but draws thunderous applause. Visible now for the first time in the play, the prime minister appears as the creation of his spin master. An item of news included in the speech hints at just how far his spin master will go to keep his creation in power: Liz has been found dead in a wing of the hotel that was engulfed in fire during the riot.

The satiric form of *Feelgood* permits expression of the darkest suspicions about the nature of contemporary political leadership while keeping the tone light. Its non-realistic situations serve as a comic medium for underscoring dismay about the actual directions of politics in Britain. The play's title, based on the song selected as the theme of the party conference, sounds a warning note about the empty reassurance of slogans manufactured by political parties. Among its comically exaggerated incidents and hilarious one-liners, the play offers a serious explanation for the current mood of political disengagement and sounds an alarm about the ultimate effect of a government that lacks substance beneath its glossy image. Arguing with Eddie, who points to the party's popularity, Liz concedes that the public is not desperate enough to turn back to the other major party, but observes that 'underneath it all, they're getting hacked off about politics' and predicts that 'they'll end up not caring who's in power. Which is why you lot aren't just going to fuck the party. You're going to fuck democracy' (75).

The Blair government's involvement in the American invasion of Iraq has steadily undermined support for its leadership. Theatre, as *Guardian* critic Michael Billington points out, has become a 'vital focus for opposition' to the war in Iraq ('Drama Out of a Crisis' 10 April 2003). In early

2003, with polls showing a majority of British citizens opposed to the Iraq invasion,[7] the first play attacking the American-led war appeared. *The Madness of George Dubya* by Justin Butcher was first performed at Theatro Technis, a London fringe venue. It transferred to the Pleasance, another London fringe theatre, and then to the Arts Theatre in the West End, where it played to sold-out houses for several months. A parody of Stanley Kubrick's anti-war film *Dr. Strangelove*, the show included updated songs by Tom Lehrer and featured daily rewrites to keep the jokes topical. It portrays George Bush as a teddy bear–clutching child who wails about weapons of mass distraction, and confuses Baghdad with Belfast. In his dream, which frames the action, a nuclear strike is launched on an Arab country called Iraqistan. The plane carrying the bomb, piloted by a couple of gay fliers, takes off from a US military base in Yorkshire, which has meanwhile been infiltrated by a female suicide bomber. Prime Minister Blair has the power to abort the bombing mission, but nearly fails to do so because he is absorbed in a real-estate transaction. In broadly satiric style, the show presents the American military establishment as rabid aggressors, insane zealots, and cynical manipulators interested in money and oil, while the American president is a puerile fantasist. The show's instructions to audience members to leave mobile phones switched on in case of a terrorist attack, and the musical finale 'We Will All Go Together When We Go', audaciously defy fears of terrorism. Propelled by the energy of its extremes, the show reveals, encourages, and exploits opposition to the war.

Distrust of government leaders takes a sharp and realistic focus in Richard Norton-Taylor's tribunal play *Justifying War: Scenes from the Hutton Inquiry*, produced at the Tricycle Theatre in November 2003. The inquiry on which this play is based dealt with very recent events that came out of the invasion of Iraq led by American and British forces beginning in March 2003. Doubts about the necessity of the Iraq invasion had been expressed from the beginning, and these doubts gained legitimacy from reports in June 2003, questioning the actual threat posed by Saddam Hussein. Specifically, the BBC reported that an unnamed expert in biological and chemical weapons familiar with the situation in Iraq had stated that an intelligence dossier about Iraqi weapons had been altered to exaggerate the threat. The alterations were said to have originated in the office of the prime minister, particularly with its Director of Communications Alistair Campbell. The government went on the defensive, and the Ministry of Defence named David Kelly, a government-employed scientist who had worked as a UN weapons inspector in Iraq, as the source of this allegation. Dr Kelly was called to testify before two parliamentary committees investigating the war in Iraq. Two days after his appearance, he was found dead. Though his death appeared to be suicide, rumours that he had been murdered to silence him began to circulate. The Hutton Inquiry, convened within two weeks of Kelly's death, was charged with examining its circumstances. It heard testimony from numerous high-level officials and government leaders, including Prime Minister Blair. The chair of the inquiry concluded that Kelly had committed suicide and found no convincing evidence of government wrongdoing (www.the-hutton-inquiry.org.uk/).

The play reveals several levels of conflict and raises far-reaching questions about the veracity and accountability of the government. The question overshadowing the proceedings is whether the government was justified in going to war in Iraq, or whether they did so on the basis of a hidden agenda and used flimsy or doctored evidence to convince a sceptical public that it was necessary. The sharpest conflict arises between two BBC reporters and the government's Director of Communication, who often clashed with the press over issues affecting the government's image. This conflict, in which it becomes clear that the government regarded the BBC as an enemy, goes to the heart of the relationship between a democratic government and a free press. The hearings also provide a glimpse of rivalries between different journalists, even those in the same organization. David Kelly nearly gets lost amid these confrontations, and it appears that both the BBC and the government used him at different times without regard for his personal position. He, of course, cannot speak for himself. His wife takes the stand on his behalf, but provides little information about his professional activities.

The phrase that recurs throughout the inquiry is 'forty-five minutes'. The dossier in question, described as 'the most substantial statement of the government's case against Iraq' (18), asserts that 'some weapons of mass destruction' could be ready to fire 'within forty-five minutes' of a decision to use them (*ibid.*). The BBC journalist reported, and maintains on the witness stand, that Kelly expressed doubt about the 45-minute time frame and indicated that the government used this unsubstantiated claim to make their case for war seem stronger. Various agents of the government assert that the 45-minute claim was made in good faith and on the basis of valid intelligence. They accuse the BBC of irresponsible reporting. The inquiry provides a platform for these opposing claims to truth, but does not affirm the truth of either.[8]

The play makes an important point, however, by concluding the section in which these opposing views are presented with the testimony of Dr Brian Jones, a retired government scientist who worked with David

Kelly on the dossier. In dry and measured sentences, Jones points out that the phrase 'weapons of mass destruction' has become a 'convenient catch-all which . . . can at times confuse discussion of the subject' (79). He distinguishes between nuclear, biological, and chemical weapons. He also reveals that several staff scientists questioned assessment of the available intelligence and considered the 45-minute claim dubious. Using precise and non-accusatory language, he states that among those working on the dossier there was 'an impression that there was an influence from outside the intelligence community' and 'an impression that they [No. 10; i.e., the Prime Minister's office] were involved in some way' (81–82).

In the several years since the Hutton Inquiry and the play *Justifying War*, events in Iraq have provided answers to some of the questions raised. Despite capture of key figures of the Saddam Hussein regime and searches of Iraq by thousands of specialists, no weapons of mass destruction (WMD) were found. No evidence of production or even the capacity for producing such weapons has been found. While the absence of WMD has become grist for jokes by comedians and cartoonists, the public has become increasingly aware of unrealistic expectations, faulty decision-making, and disastrous mistakes in Iraq. What remains relevant about the play and the inquiry it dramatizes are the questions of what kind of threat justifies war and what kind of democratic leader treats the press as its enemy.

The next examination of political leadership, following hard on the heels of *Justifying War*, also placed questions about political leadership within the context of a more specific issue. *The Permanent Way* by David Hare, a co-production between Out of Joint and the National Theatre first performed in 2003, addressed the series of five terrible train accidents between 1997 and 2002 that killed 66 people and caused hundreds of injuries. When the first accident, a collision between a commuter train en route to London and an empty train at Southall, occurred in 1997, the seven deaths were mourned, but not blamed on anything other than misfortune. In 1999, when a two-train collision in Ladbroke Grove killed 31, the public began to raise questions about changes in the ownership and management of the rail system that took effect in 1997. Pushed through by the Conservatives as they anticipated losing the coming election, the process of privatization broke British Rail into three independently operated components: rolling stock companies that owned the equipment, passenger service franchises that leased and operated the trains, and a track management company. Railtrack, the track management company, subcontracted work to regional maintenance companies.

The sale of these assets had generated huge profits for private investors, but no funds had been reserved for system maintenance or development. Despite clear evidence that the Southall and Ladbroke Grove crashes should have been prevented, no corrective measures were instituted, and three more accidents – Hatfield in 2000, King's Cross in 2001, and Potters Bar in 2002 – brought additional deaths and injuries. Following the Potters Bar crash, rail operations were severely disrupted. Fearful commuters avoided using trains. Official inquiries found pervasive mismanagement, insufficient funds for maintenance, lack of oversight, and little coordination among the separate companies.

The play's subtitle, 'La Voie Anglaise', means the English way, and the first scene shows English commuters behaving in archetypical English fashion. Waiting for an overdue train, they read newspapers and try to stifle their impatience. Their voiced thoughts, however, express a host of discontents: incompetent management of the transportation system, deterioration of public services, the loss of practical skills in the populace, the cynicism of those who profited from rail privatization, and politicians' unfulfilled promises to improve services. Several people speculate about being killed or injured in a train crash. One notes the sharp rise in government subsidy of the railways since privatization. A man asks himself, 'Why aren't people angry?' He answers his own question with the thought 'Nobody believes that by being angry, expressing anger, anything changes, anything *can* change' (9).

The play explores the railroad crisis through interviews with individuals involved in the rail system and impacted by its accidents and failures. The 9-actor ensemble of the original production performed over 50 roles in collage-like scenes that move through multiple viewpoints and incidents with locomotive speed. Monologues taken verbatim from interviews and public records describe the decline of what was once British Rail. A succession of occasionally recognizable characters explain their roles in privatization, beginning with a high-level treasury official who disdains the subject of railroads, but stipulates that his comments not be attributed to him. An investment banker, a civil servant, a rail executive, and a rail engineer provide pieces of the story. They relate that British Rail was considered a 'basket case' (10), and was sold for less than £4 per share, but then quadrupled in value within two years. Meanwhile, ridership grew enormously with modest improvements in service. Authorities ignored evidence of a serious error in separating train operation from ownership and maintenance of the tracks until the crashes occurred. A rail engineer criticizes the replacement of engineers with managers. A group of line workers describe a culture of barely trained and poorly supervised hourly labourers with no job security. The picture emerges of a system dedicated to rewarding investors rather than serving passengers.

Individuals directly impacted by the accidents speak about their losses, creating the most powerful moments in the play. Two women whose adult children died in the crashes tell of confusion and mismanagement following the accidents. The first learned of her son's death from a television broadcast. Another went to a station for information and was notified of her son's death in this public, crowded place. Both mothers emphasize not just their grief, but also their outrage at getting nowhere in their quest for accountability, for rail safety improvements, and even for respect. Those who were injured but survived rail accidents describe moments of horror followed by years of attempting to return to normality. A transport police officer reveals that his attempts to investigate were obstructed. Some of the families and survivors have organized to demand changes, but a split has developed between those who lost loved ones and those who were injured. Individual speeches merge into a collective narrative and chorus of emotions punctuated by the metallic noise of trains. The voices cease only in the climactic moment when a deafening clash of steel on steel obliterates all other sounds.

Loss of identity and questioning of values emerge as the primary themes. Except for the government ministers who conducted the official inquiry into the accidents, speakers are not named. Erasure of personal identity appears as both a contributing factor and a result of the accidents. Anonymity and lack of continuity removed accountability, as managers were shuffled from one organization to another and rail workers were known only by the names on hand-me-down hard hats. In the wake of the crashes, accident victims experienced a loss of identity that compounded their pain. A bereaved mother tells of her outrage over the reference to her son as 'Body No. 6' in the accident report (27). Another sobs as she recalls being asked to estimate the monetary value of Christmas presents from her son in order to establish his VOL (Value of Life) for compensation. The bereaved cannot forgive the railroad companies for placing profit above human life. A woman who suffered injuries and lost her husband in the Potters Bar derailment speaks bitterly of a fundamentally corrupt society in which politicians and rail officials advance their own interests at the expense of the public. Survivors implicitly and sometimes explicitly express deep disappointment in government officials and political leaders who have abandoned responsibility and honesty.

The playwright's identity interacts with the play's structure in an unusual form of communication and a scope of inquiry that opens out beyond the railroad system. Speakers occasionally say David Hare's first name as though conversing with him. This reference not only keeps the notion of documentary-style reporting in the foreground, but also reminds the audience that despite its factual base and collaborative creation, the work expresses Hare's viewpoint. One of Britain's most visible and respected playwrights, Hare invokes his personal experience and position when he places responsibility for the rail disasters upon both major political parties and all corporate entities involved in the railroads. At the same time, the play's episodic structure combines with the effective doubling of roles to create the vivid impression of a society in which responsibility has been atomized and dispersed to the point where blame is futile and change unlikely. Hare ultimately identifies a tragedy even larger than the cumulative losses of the accidents or the failure to maintain the rail transport system. That tragedy is the failure of governmental leadership and accountability in a nation that has not only referred to its rail system as 'the permanent way', but also considered its political and social institutions as an exemplary, effective, and permanent way of conducting public life. This failure prompts even more urgent consideration of the question asked repeatedly throughout the play: why aren't people angry?

Performances of *The Permanent Way*, in fact, seemed to tap into a reservoir of public anger. The piece played to sold-out houses in the National's Cottesloe, on tour, and in the larger Lyttleton Theatre when it returned to the National. Though public attention had been directed to the series of crashes and the demonstrated failures of railway lines and safety equipment that caused them,[9] the play served to galvanize public opinion on the issue. Audiences were visibly moved by the stories of individual loss, survival, and courage, and shocked by the evidence of mismanagement and disregard for safety. Survivors who attended the play's opening at the National praised it. Well-known novelist Nina Bawden, who lost her husband and sustained serious injuries in the Potters Bar crash, stated, 'It would be nice if all the Government were taken in manacles to it' (*The Observer* 18 January 2004, online edition). Rail industry officials criticized the play, claiming it misrepresented the facts. Two months after its opening, Transport Secretary Alistair Darling made a speech to the House of Commons acknowledging the 'history of under investment' in the railway system, the 'poor state of much of the railway infrastructure', and the 'fragmentation, excessive complication and dysfunctionality' that resulted from privatization ('Full Text: Statement on the Railways' *Guardian* 19 January 2004, online edition). The speech also pointed to increased public funding and the replacement of Railtrack with Network Rail, which had agreed not to subcontract maintenance, but criticized 'over-cautious' safety regulations (*ibid.*). Turning over the rail system to Network Rail effectively renationalized that component of the system, but Jarvis, a company that was found negligent in the Potters Bar crash, continued to hold contracts for maintenance. Investigations of the accident at Potters Bar continue. Some of the bereaved campaigned to have individuals charged with manslaughter, but that effort did not succeed, and since 2004 the rail safety issue has receded from public view.

David Hare continued his examination of political leadership with *Stuff Happens*, produced by the National Theatre in 2004. This play examines the involvement of the Blair government in the decision to go to war in Iraq. Its title seizes on Donald Rumsfeld's dismissive explanation when asked about widespread looting in Baghdad following the overthrow of Saddam Hussein – a phrase that has come to exemplify the inadequate response to chaos in Iraq. Hare has constructed the play by combining monologues based on published sources with scenes depicting what might have happened in various private meetings. Action ranges from private meetings at the White House or the Bush ranch to debates in the British Parliament, the American Congress, and the United Nations. The more than 50 characters played by 18 actors represent known members of the American and British governments involved in the decision to invade Iraq. American characters include George Bush, Vice President Dick Cheney, Secretary of State Colin Powell, advisor Condoleezza Rice, Secretary of Defense Donald Rumsfeld, CIA Director George Tenet, advisor Paul Wolfowitz, and a number of others. Representatives of the British government include Tony Blair, Foreign Secretary Jack Straw, Foreign Policy Advisor David Manning, Director of Communications Alistair Campbell, Defence Minister Geoff Hoon, and others. The French President Jacques Chirac, former Iraqi dictator Saddam Hussein, UN Inspector Hans Blix, UN Secretary General Kofi Annan, and cellist Yo Yo Ma contribute to the unfolding story. Unnamed actors narrate background, identify places and times, and introduce characters. The skilfully compiled narrative uses irony effectively, creates suspense, and introduces elements of surprise, even though the events it presents are generally known.

Stuff Happens attempts to answer the question of why the war in Iraq happened. It finds an answer in the personalities and goals of major leaders in the West. Inevitably, the relationship between George Bush and Tony Blair assumes central importance, but, in Shakespearian style, subordinate figures appear first and verbally build the scenario. Colin

Powell voices a belief, which he says grew out of the experience of Vietnam, that 'war should be the politics of last resort' (5). Dick Cheney explains that 'other priorities' prevented him from serving in the military during Vietnam, while a memo he issued in 1974, about a malfunctioning drain in a White House sink, mocks this explanation (5–6). Brief descriptions depict Rumsfeld as aggressive in the extreme, Condoleezza Rice as irresolute and mistrustful, and advisor Paul Wolfowitz as ultra hawkish. Tony Blair, described as enigmatic except for his ambition, is introduced along with Kofi Annan, who radiates an air of mysticism, and Swedish UN weapons inspector Hans Blix, whom Powell calls 'reliable as a Volvo' (9). Bush introduces himself, opening with the statement, 'My faith frees me' (9). Narration informs the audience of Bush's background of heavy drinking and unimpressive performance, which he apparently transcended through religion.

   The build-up to war revolves around George Bush's unhesitating grasp of power and his focus on Iraq from the start of his presidency. Meeting with his cabinet and advisors ten days after taking office, Bush focuses on indistinct CIA photos purporting to show weapon development in Iraq. The attack on the World Trade Center brings his declaration of war, followed by an executive order to attack Afghanistan. Bush parades Tony Blair before a cheering US Congress, but soon fails to reciprocate Blair's support. When Blair asks Bush why British special forces had suddenly been ordered to pull back when they were close to capturing Osama bin Laden, Bush displays bland indifference, offering no explanation or apology. Bush appears to listen more than he talks, but often cuts off discussion with a blunt reminder of his power. He sometimes speaks indefinitely but never appears to swerve from the intention to invade Iraq and overthrow Saddam Hussein. This single-minded determination, which admits of no uncertainty, submerges Bush's weaknesses of character and leadership in its strength and sweeps aside opposition. In the statement 'I'm a war president' (49) Bush finds the authority he desires.

   Blair's caution, uncertainty, and desire to appear independent make him seem weak, in contrast to the steamroller force of the American president who believes he is acting on behalf of God and commands the superpower's financial and military might. During a visit to Bush's Texas ranch, Blair urges an even-handed approach to the conflict between Israel and the Palestinians, but Bush only wants to talk about Iraq. Concerned with the political implications of growing opposition to a US-led war in Iraq, Blair insists on support of the United Nations and evidence of a 'real and imminent' Iraqi threat (39). Blair asks Bush for assurance that no decision has yet been made, emphasizing the importance of good faith.

Bush gives that assurance, but the lack of precision in his speech leaves Blair unsure of what has been agreed. In an attempt to protect his position, Blair orders an intelligence dossier assembled on the Iraqi threat. This dossier's claim that Iraq could launch chemical or biological weapons within 45 minutes would become central to the controversies surrounding the death of David Kelly. The play shows Blair unconvinced of the validity of this intelligence but desperate to rationalize a decision that has already been made for him. Later he agonizes over the conflict between the expectation of the British public that the country will not go to war without a UN resolution and his belief that his own political future depends on the support of the American president.

The play finds a further dimension of dramatic conflict in the relations among Bush's advisors. From the beginning, Colin Powell insists on the importance of assembling and acting with a coalition. He employs a formidable array of skills when dealing with potential coalition partners, like France, who refuse to approve US actions without sharing in decision-making. As the only one in Bush's circle with military experience, Powell expects to be taken seriously when he warns against rushing into war; however, he consistently finds himself an outsider. The behaviour of advisor Condoleezza Rice, the only other African American present in the meetings, provides an ironic contrast with Powell's attempts to maintain personal integrity. In a telling moment, as cabinet members and their wives wait uncertainly for a signal of closure at the end of a long day, Rice suggests going bowling, but Laura Bush asks for a hymn to be sung. Obliging the woman whose role seems to shift subtly from boss's wife towards plantation mistress, Rice sings the hymn, while Powell remains coldly distant. Rice seems to identify so closely with Bush that she has no need for independence, but does venture to suggest that Bush inform Powell when he remains the only member of the inner circle unaware that the decision to go to war has been made. Powell nevertheless maintains his attempts to negotiate until all options are exhausted, giving his word again and again. Repeatedly, Rumsfeld and Cheney undermine Powell's work, discrediting negotiations and disregarding assurances he has given.

The fates of Powell and Blair converge at the climax of a feverish campaign to pass a UN resolution that unequivocally authorizes the invasion of Iraq. As Blair, who has promised not to go to war without such an authorization, gives his utmost in both public and private efforts to secure passage, the Bush administration decides that Blair is expendable. Powell expresses outrage at this betrayal of a friend who has been 'loyal from the start' (104). Cheney insults Powell, signalling that he, too, is expendable. Subsequently, Rumsfeld states in a news conference that Britain is not essential to the plans for war, placing Blair in an untenable position. Powell accepts the unacceptable, taking to the UN Security Council the flimsy evidence of Iraq's WMD – evidence subsequently admitted to be invalid by the Bush administration.[10] To save himself politically, Blair seizes on a statement by the French to blame them for the breakdown of negotiations in the United Nations. Both men thus find their personal credibility undermined or destroyed by Bush's unwavering determination to go to war and by the devious manoeuvres of Cheney and Rumsfeld.

Though the play expresses and arouses anger about the way in which the Iraq war began, it touches on divergent perspectives in relation to this war. The Member of Parliament highlighted in Act II, who claims that 'Bush will hit Iraq in much the same way that a drunk will hit a bottle – to satisfy his thirst for power and oil' (78) and urges Britain to be the friend who denies the drunk that bottle, speaks for the majority of war opponents. Other opinions include that of a British journalist who approves the ouster of a tyrannical dictator, asserting that only Westerners who take their own freedom for granted would question the means of liberation. A New Labour politician speaks of the difficulty of assessing the long-range effect of the war, considering the contradictory outcomes of removing a dictator but finding no hidden weapons, and achieving an initial victory but failing to establish peace and stability. A Palestinian places the Iraq war in the context of US support of Israel. A British visitor in New York comments on the inability of Americans to accept the views of non-Americans on issues relating to terrorism. The play gives the final words to an Iraqi exile who had longed for the fall of Saddam Hussein but who is angry at the great harm done by the invaders. He predicts that Iraq will continue to suffer until it takes responsibility for itself.

This play joins numerous books and other creative works, such as the Michael Moore film *Fahrenheit 9/11* (2004) in arguing that the Bush administration misled and betrayed the United States and the world in its invasion of Iraq. After a heavily publicized, much-discussed, and sold-out initial run at the National Theatre, it was inexplicably taken out of the National's repertory and remained unperformed until American productions in the summer of 2005 by the Mark Taper Forum in Los Angeles, and in early 2006 at New York's Public Theater. In revisiting the beginning of the war, the play reveals Bush administration malfeasance that has since become widely known.[11] It speculates that Blair committed Britain to the war because he had learned from seeing his Labour predecessor Neil Kinnock lose after being brushed off by American President Ronald Reagan that power in Whitehall depends on support from the White House. The play does not explore the possibility or implications of separating British interests from American politics, or what steps Britain might take to disengage from the war. It focuses completely on making visible the irresponsible, self-serving, and irrational qualities of current political leaders. Its context-specific study of political leadership contains important implications for future leaders and political processes.

David Edgar examines contemporary political leadership in the context of regional politics in *Playing with Fire*, produced by the National Theatre in 2005. This play bases its action on New Labour's attempts to bring local councils into compliance with new standards regarding collection of taxes, provision of services, educational outcomes, programming to combat racism, and communication with the public. Set in the fictional West Yorkshire city of Wyverdale, the play unfolds in a mixture of narration and scenes. The deputy prime minister sends civil servant Alex Clifton to Wyverdale to initiate and manage changes in the way its District Council operates. He describes Alex as Wyverdale's 'Poland' solution – that is, a chance to adopt the policies dictated to them – and also invokes the threat of a 'Czechoslovakia' solution in which local government would be taken over. Alex, a divorced woman in her late thirties, has not only ambition and a brisk style, but also a streak of romantic idealism. When she arrives in Wyverdale, she walks the business district and stops in a beauty shop for a manicure, encountering at first hand the city's poverty, lack of education, and ethnic division.

The Wyverdale Council welcomes Alex with a somewhat chaotic reception. In conversations constantly interrupted by ringing mobiles, she meets the members. George, the council leader, a middle-aged, working-class man with old Labour sympathies and habits, feels that central government officials never see 'Cool Britannia's Back Side': the 'car parks, car graveyards, gardens, ginnels and what's dumped in 'em' because they travel in cars rather than trains (8). Arthur, an old-school local politician, confronts Alex with gruff resentment, while Frank, the Bible-quoting former headmaster who is Cabinet Member for Education, attacks with more subtlety. Riaz, an Asian member of the Council, offers the only sign of friendliness. To impress Alex, each member has been primed to deliver a bit of information, demonstrating success in matters such as police patrols, park improvements, and number of children in care. When she asks about problems, they bring up prostitution and challenge her to prove that she can accomplish anything. Taking the dare, she goes out after the meeting, accompanied by Riaz, speaks with a young prostitute, and proposes a drop-in centre for drug rehabilitation and job training.

The Council members drag their feet until a second consultant arrives. Her session of motivational training with a heavy dose of acronym-laden corporate speak bullies the Council into making changes. These changes, however, lead to unforeseen consequences. To finance a website, multilingual translations of information brochures, training for councillors, and multicultural festivals, the Council closes down swimming facilities and children's homes. Contracting out some services saves money but increases local unemployment. The Council wins a European Union grant to upgrade housing on an estate with a primarily Asian population, but must provide matching funds, and therefore suspends routine repairs at another estate populated primarily by whites. Tensions escalate, and a white youth is killed in an altercation between racial groups. The rising hostility strengthens the far-right Britannia Party, and they disrupt a Holocaust commemoration. Frank, who was forced to turn over his cabinet position to Riaz so that the cabinet would not be all-white, resigns from the Labour Party, denounces the reforms, and announces that he will henceforth sit as an independent.

The second act begins with a tribunal-like play within the action, inquiring into the circumstances that led to a riot in Wyverdale several months after the white youth was killed. Both Asian and white citizens bring to this forum the resentments that have built up during the year. The changes initiated by the central government have fuelled anger and led to the hardening of racial divisions. Whites regard the housing upgrades, translated brochures, and anti-racism measures as favouritism towards Asians. Asians think the new drop-in counselling centre promotes prostitution and drug use. Though the two sides seem far apart, a surprising moment of convergence occurs when an Asian youth suggests that the loss of English small-town culture remembered from the past has been caused by breakdown among white families, and the mother of the youth who was killed expresses agreement.

The riots have changed the city politically. In the recent election, the Labour Party, always dominant in local politics, has been voted out of power. Frank, standing as an independent, has been elected mayor, and the Britannia Party has picked up two seats. In a concluding flashback scene that takes place during the riot, George remarks to Alex that the model for Wyverdale is neither Poland nor Czechoslovakia, but Kosovo: 'You do the right thing from the best of motives. But you do it from a height of fifteen thousand feet. Which means you hit a lot of things you didn't mean to . . .' (133). Stung by George's judgement that she would have seen things more clearly and acted differently if she had been part of the community, Alex prepares to stay permanently.

This play criticizes the leadership within Labour's ruling elite for failing to understand the complexity and interconnectedness of issues in local government. It condemns the interference of career politicians and bureaucrats in situations they have not experienced and do not understand. It challenges simplistic notions of overturning privilege and promoting cultural diversity, deriding public multicultural celebrations with artificially heterogeneous participants and overworked platitudes. While it does not endorse local government practices that preserve white privilege, the play makes the point that in a divided city, taking resources from one group and giving them to another will only harden divisions. It illustrates the limits of diversity in the moment during the tribunal hearing when the white mother wants to show agreement with the Asian youth and extends her hand to him, only to be ignored because of his religious stricture against touching a woman outside his family. In contrasting references to an Asian man who escaped censure for sexually harassing a female employee and a white man who has been stigmatized for decades because he had sex with a 19-year-old male, the play further emphasizes the conflicts inherent in incorporating different cultural traditions into one society. In turning a spotlight on the rivalry, suspicion, and threat of violence inherent in racial separation, the play questions whether such separation can be maintained peacefully. In Frank's prediction that Riaz will not be re-elected because his Bangladeshi origin differs from the Pakistani identity of most Asians in the city, the play further suggests that ethnic identification tends to promote disunity rather than community.

The ending of *Playing with Fire* presents two alternatives. The first, exemplified by Frank, involves turning away from the Labour Party, moving towards the right, and appeasing the far right in an attempt to ensure that working-class white communities and individuals are not left out of government's solutions to current problems and planning for the future. As Edgar notes in an afterword, far-right politics is resurgent in Europe, and the British National Party has made gains in recent elections. The second, exemplified by Alex, involves the decision to remain in Wyverdale for the long term, devoting herself to understanding its people and working to unify and strengthen them. Edgar depicts her as abandoning ambition in favour of idealism, as she prepares to bring young people from white and Asian backgrounds into the room where the Council has been meeting and teach them the team-building skills that the Council refused to learn. Through Alex, Edgar expresses hope that idealistic individuals with education, experience, and skills will create new forms of community and solve or move beyond the conflicts of the past. Through her example, he also appeals for reasonable approaches to countering division and extremism.

Rather than bringing little-known phenomena to public attention, the issue plays interpret events or developments that have already captured public attention. Their interpretations usually begin with questions, such as why the murderers of Stephen Lawrence have not been brought to justice, why Britain became involved in the Iraq war, or why race riots have occurred during a time when the government appears to be making efforts to improve conditions for non-white communities. Most of the plays address the question or issue through open-ended analysis rather than by arguing for a particular answer or solution. They frame the analysis in terms of universal values such as rationality, justice, dignity, and peace rather than in terms that express the values of particular parties or social groups. Some include points of view that seem to clash with the presumed politics of the playwright. The interpretive thrust shared by the issue plays subjects to interrogation the information and forms of communication made available by official sources, including the government and the media.

Issue plays of the post-Thatcher era show significant differences from those of the 1970s and 1980s. The use of documented evidence stands out as the most important innovative feature in this group of plays. The tribunal plays, as well as those that rely in part on verbatim interviews and published material, serve to educate the audience about the context for the issue while presenting it. The plays that do rely on shared knowledge or experience base their action on currently unfolding events such as the peace process in Northern Ireland, the tensions within New Labour, or the war in Iraq. Headline-oriented topicality and in-jokes (about, e.g., being able to tell the difference between mushy peas and guacamole) abound in the plays that rely on shared knowledge, lending them the authority of recognition.

The issue that most emphatically and consistently emerges in these plays is that of political leadership. Intriguingly, the issue plays include no analyses or appraisals of the Thatcher government.[12] They address the current situation, registering dissatisfaction, unease, or anger with regard to aspects of post-Thatcher society and the leaders who have shaped it. The vigorous critiques, coupled with illustrations of the inadequacy of existing formulae for addressing complex and intransigent problems, point to the necessity for new political approaches as well as new leadership. Although they shine a bright light on the failures of existing leaders, these plays barely hint at new approaches and offer no positive models to counter the negative portrayals of the current leadership. Rather than envisioning ideal leaders, imagining utopias, protesting the establishment, or advocating specific changes, these issue plays are intent on exploring and exposing the contradictions and failures of a turbulent present.

6
Post-Thatcher Britain and Global Politics

The scope of British political theatre expanded beyond Britain's borders in the latter part of the twentieth century. During the 1960s and 1970s, London provided an artistic home for political playwrights such as Wole Soyinka and Athol Fugard, who were exiled from or unable to secure productions of their plays in their home countries. From the early 1980s, British theatres regularly showcased foreign companies through international festivals and events. Most recently, British playwrights have begun to address aspects of global politics. International themes arise from a consciousness that Britain's current problems and opportunities arise in the context of multiple connections with other nations. The global issues of concern to post-Thatcher playwrights include wars and conflicts, refugees, transnational corporations, and migration in and out of Great Britain. Plays that assert Britain's political stake in situations beyond its borders often use factual materials, including official tribunals, interviews, and published sources, but may frame such material in a personal viewpoint. The personal perspective may make it clear that the writer is an outsider to the conflict, or may heighten the urgency of the issues, particularly in plays about asylum seekers in England. Some plays combine the outsider position and the sense of moral urgency to signal uncertainty regarding the truth of a situation and the best national response to it.

Conflicts and wars

Concern about war pervades post-Thatcher British drama, taking a variety of forms. A number of plays, including Caryl Churchill's *Far Away*, Richard Bean's *The Mentalists*, and Bryony Lavery's *Frozen*, use abstract or metaphorical constructs to evoke war. Other contemporary plays, such as *East Is*

East by Ayub Khan-Din and *Copenhagen* by Michael Frayn, use war as a context shaping actions and outcomes. Still others, such as David Hare's *Stuff Happens*, use war to expose failures of political leadership. The plays discussed in this section, however, deal directly with recent or current conflicts and wars, approaching them as political issues in themselves. Recent plays highlight four areas of conflict: the Bosnian–Serb–Croat war in the former Yugoslavia, the ongoing violence between Israelis and Palestinians, the ethnic wars in Africa, and the Iraq war.

The conflict that broke out in the former Yugoslavia in 1992 introduced contemporary Europe to the horrors of ethnic cleansing: mass killings and systematic rapes aimed at eliminating one group from a region so that a rival group could take possession. Thousands died before nations in the NATO alliance accepted the reality of ethnic violence and agreed to intervene. The United Nations established safe areas that included Srebrenica in 1993, but UN forces were unable to prevent the Serb army moving into Srebrenica in July 1995 and killing 6000–8000 Bosnian Muslims. Only after reinforcements and new leadership arrived were UN forces able to halt the violence.

Srebrenica by Nicolas Kent, produced by the Tricycle Theatre in 1998, presents edited transcripts of the war crimes tribunal begun in 1996 to investigate the actions of the Serb army in Srebrenica in the summer of 1995. This work deals with both the war crimes perpetrated by the Serbian army and the failure of the UN forces to prevent the massacre. It includes testimony from the commander of the Dutch peacekeeping force acting on behalf of the United Nations, a member of the Dutch forces, and a member of the Bosnian Serb army. Witnesses describe atrocities that range from sadistic cruelty towards individuals, such as forcing a father to eat flesh from his disembowelled child, to mass shootings of truckloads of men in a deserted field. UN officials describe a terrorized populace frenziedly seeking some sort of safety and an overwhelmed and ineffective protecting force.

As one of the first tribunal plays, *Srebrenica* lacks some of the qualities that made the later ones successful as theatre. Because it excerpts long sections from the transcripts, the piece takes the form of a sequence of speeches rather than dialogue. It does not include any direct or written testimony from survivors of the massacre, and thus does not give them visibility as specific individuals. Most statements come from the members of the UN force, who were third-party participants attempting to protect the people of Srebrenica, and they focus more on their attempts to manage the crisis than on the crimes that occurred beyond their control. The Serbian army witness who confesses to participating in the execution of unarmed civilians describes himself as an outsider, condemns the actions of the army, and claims that if he had refused to participate he would have been killed. The play uses testimony to establish that mass killing occurred, but lacks the dramatic tension of a confrontation between the perpetrators and the victims. Nevertheless, it establishes the Srebrenica massacre as an international political issue. Its vivid descriptions of violence against civilians make visible the moral outrage of ethnic cleansing and pose the question of how this deadly conflict in nearby areas of Europe overran international efforts at peacekeeping and protection of civilians. The many unanswered questions evident at the end of the play deny closure and place in doubt the possibility of justice for the victims of the conflict. Perhaps the play's most salient political point lies in exposing the fragility of institutions charged with the responsibility of peacekeeping while emphasizing their vital importance. The savagery of the violence in a European enclave such as Srebrenica shocked the British public. This play revisits that sense of shock, while leaving unsettled the question of the capacity of a wider European community to protect civilians.

David Hare created a highly personal examination of contemporary conflict with his *Via Dolorosa*, written after a visit to the Middle East in which he interviewed Palestinians and Israelis to learn about their experiences and perspectives. The Royal Court first produced *Via Dolorosa* in 1998. Hare wrote the piece as a first-person monologue describing his interactions with Israelis and Palestinians, and performed it himself.[1] Hare begins by exposing his own position as an outsider to the conflict – a man from England, a country where 'no one believes in anything any more' (4). Troubled by his unbelief, he states that for some time his writing has reflected a preoccupation with faith. He sees Israel as 'first and foremost . . . a cause' (*ibid.*), and identifies that cause with the survival of the Jewish people. An Israeli novelist who visits him in London before the trip argues that the Six-Day War created a preoccupation with the ownership of land that has proved destructive to the ideas manifested in Israel's founding. This opposition between land and ideas comes to dominate Hare's view of the conflict. When he interviews Israelis, Hare soon discovers the multiple divisions among them, such as the split between secular and orthodox Jews, and between those who live inside Israel's internationally recognized borders and those who have settled on land captured in 1967. Visiting a settlement, he meets a family devoted to a belief in the historical right of Jews to the Biblical lands of Israel. Disagreement erupts among this family, too, first in a heated argument about the 1995 assassination of Yitzhak Rabin and later in a spirited debate over a detail in a Bible story. In Tel Aviv he listens to a socialist former associate of Rabin express extreme anger and despair at Israel's current political choices.

Crossing into Gaza, Hare is struck by the contrast to Israel, comparing the transit to 'moving from California into Bangladesh' (24). He describes piles of garbage and debris and quotes statistics on the dire poverty of the Palestinians. Interviewees include a political leader who criticizes PLO corruption and asserts that Palestinians need 'to reform ourselves' (27) and a woman working to establish a civil service who thinks the *intifada* has drained the Palestinians of energy for nation building. A historian in Ramallah deplores the everyday indignities forced on Palestinians by Israeli security but feels that Israelis and Palestinians are 'bound up in each other's unhappiness' and 'cannot be separated' (31). A theatre director describes a cross-border production of *Romeo and Juliet* that ended in bad feelings. A Palestinian intellectual decries the destruction of non-Western cultures by the United States and Europe.

Having sampled both sides of the conflict, Hare visits Jerusalem, the 'world capital of claim and counter-claim... the acknowledged metropolis of dispute' (36). There he attempts to experience Christianity by walking the Via Dolorosa, or Stations of the Cross. This pilgrimage fills him with 'a sense of loss' (*ibid.*), and the Church of the Holy Sepulchre at its end proves to be yet another site of division, dispute, and uncertainty. Seeing Christians kneeling and kissing a stone that may mark the site of the crucifixion, Hare echoes the words of the Israeli novelist when he questions whether they are kissing a stone or an idea. He visits the Holocaust Memorial, where he finds the works of art inadequate, compared to the impact of open shelves of files documenting millions of atrocities and deaths.

Hare ends the trip thoughtful and troubled, but unable to integrate what he has learned in the Middle East with his own politics. He details his return to the familiarity and security of home, describing the route from Victoria Station and ending with the 'white door' of his house closing (43) to signal a definitive separation between his personal world and that of Israel and Gaza. Remembered images and fragmentary speeches resolve into the simple opposition 'Stones or ideas?' (*ibid.*). The cryptic coda 'Via Dolorosa' evokes a fated suffering that is outside politics and even outside history. It gives audience members an opportunity to empathize with people caught in the conflict, but does not suggest a constructive means of engaging with the difficult, complex, and inevitably painful process of resolving it.

Addressing an issue with passionate partisans in Britain and the United States, *Via Dolorosa* aroused considerable controversy. Both Palestinians and Jews attacked the play in post-performance discussions and in print, especially when it was performed in New York in 1999. Some of those who were interviewed claimed that their statements had been misrepresented. Hare replied in a preface to excerpts from the piece published in the *Guardian* in 2000. When the show was revived in England in 2002, the well-known Jewish writer Arnold Wesker published an open letter admonishing Hare for political correctness in blaming Israel for the suffering of Palestinians. *Via Dolorosa* added to the visibility of the Israeli–Palestinian conflict, stimulated debate, and to some extent brought into the debate those without ties to either side. It demonstrated both the potential and the limits of an individual approach to an international dilemma.

A no-less personal but very different perspective on the Palestinian–Israeli conflict appears in *My Name Is Rachel Corrie*, produced by the Royal Court in 2005. This one-woman play presents the activist life and passionate views of the young American woman killed on 16 March 2003 by an Israeli army bulldozer as she was trying to defend a Palestinian home from demolition. A verbatim piece based on Rachel Corrie's diaries, letters, emails, and other writings, it was edited by Katherine Viner, a journalist, and Alan Rickman, an actor who also directed the play. Presenting the young woman's perspective in her own words, the play sidesteps the disagreements about her action and Israeli responsibility. The play generated controversy of its own, however, when a New York run, scheduled by the Theatre Workshop, was postponed indefinitely in response to a storm of protest about a perceived anti-Israel bias of the play. That move brought strong protests from playwrights, actors, and other advocates of the freedom to air controversial issues in the theatre. The piece then played in London's Playhouse Theatre and at the Edinburgh Fringe Festival prior to performance in Seattle and New York in 2006.

The narrative begins early in Rachel's life. She grows up in Olympia, Washington, surrounded by the natural beauty of the Pacific Northwest. Her activism emerges at a young age as she takes the podium in a fifth-grade programme, urging the eradication of hunger by the year 2000. Visiting Russia at the age of 15, she describes her host city as 'flawed, dirty, broken, and gorgeous' (7). Rachel hopes for a more meaningful life than she observes among the apathetic former hippies of her parents' generation. Asked what she wants to be when she grows up, she writes a 'five-paragraph manifesto on the million things I wanted to be' (6). Rachel dreams of 'building the world myself and putting new hats on everybody' (*ibid.*). At Evergreen State College, her idealism propels her into the whirl of activity described in a diary entry:

I'm still pretty shell-shocked by this semester. I spoke to a room of about forty international students. I've helped in the planning of two conferences, facilitated meetings, danced down the street with forty people from the ages of seven to seventy, dressed as doves. I spent a lot of time with the homeless group. I went with them to city council. I went to the community conversation. I slept out overnight on May Day. (11)

Rachel decides to travel to Gaza, despite her parents' fears for her safety. She joins the non-violent International Solidarity Movement, which uses American and other foreign volunteers to shield homes and community resources from destruction by the Israeli army. The play follows her up to the day of her death, as she quickly connects with people in Gaza but realizes she has much to learn about their history and culture.

Katherine Viner, one of the play's editors, writes in an afterword to the published edition, 'we wanted to uncover the young woman behind the political symbol' (54). The character that emerges through Rachel's own words is a genuine individual, not an icon. She has bad habits, like smoking and losing pens. She suffers personal hurt when a boyfriend leaves her, as well as the more abstract hurt from the realization of 'how awful we can allow the world to be' (49). Her perspective mixes questioning and certainty, realism and fantasy in ways typical of young people. At the same time, she shows a commitment to action that is far from typical. Many people voice concern about conflicts in places far from their home, but even the few who visit to take a close look are unlikely to commit themselves to risky forms of intervention. Rachel Corrie risked and ultimately lost her life in the cause of protecting people she felt were innocent victims of aggression. She likely did not understand all the implications of her actions, since she had been in Gaza a bare three months. Her sacrifice did not save the home she was trying to protect or change the tactics of the Israeli army. It does, however, make visible the possibility of demanding more of the world.

The play extends the impact of Rachel's sacrifice both in its affectionate portrait of her as a likeable individual and in its serious attention to her remarkable moral vision. She forcefully articulates the cause she represented, insisting that the audience take notice when Palestinians are killed, when the greenhouses that provide their livelihood are destroyed, and when even young children are threatened by an omnipresent military. What she sees in Gaza calls forth Rachel's outrage. Since second grade, she writes, she has maintained firm belief in the precept that 'everyone should feel safe' (5). She now sees evil in a global political structure that denies safety to children in Gaza and protests that 'this is not at all what I asked for when I came into this world' (50).

My Name Is Rachel Corrie and *Via Dolorosa* use similar artistic strategies but show starkly opposing political attitudes that epitomize the post-Thatcher generational divide. Hare's piece offers the perspective of an accomplished and respected man who has seen much of life and the world. It offers carefully balanced glimpses of people and opinions on both sides of the divide. Hare ultimately returns to his safe zone, closes the door, and places the conflict within a tragic paradigm evoking the suffering of Christ. His writing and performance of the piece occurred in the continuum of a long and distinguished career. *My Name Is Rachel Corrie* communicates the perspective of a young and unknown woman who wholeheartedly devoted herself to the welfare of Palestinians whom she hardly knew but championed because they needed help. Attempting to lend them the protection she enjoyed as a privileged American, she sacrificed her life and with it all individual future potential for learning more about the situation, changing her mind, or walking away from the conflict. The play situates the meaning of Rachel's act in her example of selfless courage, but contests a view of her death as a fated tragedy, emphasizing the absence of closure. Rachel herself is absent from the final process of constructing meaning. Both her absence and Hare's presence raise the questions about the legitimacy and effectiveness of individual action in foreign conflicts. Such questions form the dominant element in post-Thatcher plays that address such conflicts.

Regional conflicts extend beyond their borders in a number of ways – not least, through the refugees forced to flee them. In the post-Thatcher years, survivors of various foreign conflicts have found refuge in Britain and, to an extent, have brought the issues and moral dilemmas of these conflicts with them. *Sanctuary* by Tanika Gupta, produced in 2002 by the National Theatre, constructs its action around a small but diverse group of people who seek escape from the violent conflicts of their past in a London churchyard. Written in realistic form, the play presents vivid characters, but its contrived plot strains the credibility of contemporary audiences. The play's setting creates a paradox: it is a graveyard, but at the same time teems with life, because the church's caretaker has created there a lush garden with a pond and resident waterfowl. Though the garden makes life visible, a threat hangs over this fragile emblem of hope: because of its dwindling and elderly congregation, the church is to be sold and converted to a health club.

Each individual in the group has found his or her way to the church-yard because of loss and a need for regeneration. Kabir, the Muslim care-taker of the church, lost his wife and homeland in the conflict over Kashmir, but his hope for a new life is evident in his garden. Michael, a Rwandan refugee and formerly a Christian pastor, silently carves wooden spoons and seems devoid of hope. Sebastian, an Afro-Caribbean photo-grapher whose career as a photojournalist was destroyed by his drinking problem, has begun to teach a children's art class at the church. Ayesha, a teenage girl, escapes a turbulent home life and studies for her exams near the grave of her father. Jenny, the priest who guides this small flock, lost her fiancé. Jenny's grandmother Margaret, who lives with her, lost her customary way of life when her husband, a colonial military officer, died. Margaret's naively insensitive comments about race, nationality, mar-riage, and sex create comic moments but indicate her fundamental estrangement from the contemporary world.

Members of this diverse group become friends and almost begin to form a community, but ultimately cannot exclude violence from their sanctuary. Kabir, desperate to exorcise his guilt about the past, confides his story to Michael. When soldiers raped and strangled his wife, he remained hidden with his infant daughter and did not try to defend her. His shame prevents him from returning home to see his daughter. Michael attempts to comfort Kabir with his story. He hid a Tutsi family, but the militia found them and forced him to kill the family and watch the slaughter of his own wife. The following day Sebastian confronts Michael with accusations that contradict his story. Producing a Bible that Michael had misplaced, he accuses him of ordering the killing of five thousand Rwandan Tutsis who had taken refuge in a church, includ-ing the Bible's original owner. Sebastian saw the carnage at the church when he was a photojournalist and has since been pursuing the man responsible. Michael denies the accusations but flees.

The consequences of these conflicts manifest themselves late that night. As Kabir digs in the churchyard, Michael appears and pleads with Kabir to save him from being returned to Rwanda, offering him a diamond. Kabir, however, has unearthed the box that Michael had buried, and in it has found Michael's passport and other evidence that validate Sebastian's accusations. Michael then admits to killing defenceless people and raping women but insists it was necessary for survival. He begs and prays, but in Kabir's mind Michael has become synonymous with the men who raped and murdered his wife. Kabir kills Michael with great purpose and vio-lence, shouting 'Now I am a Hutu and you are a Tutsi' (104) and quoting the Biblical phrase 'an eye for an eye . . .' (106).

Sunrise reveals an exhausted Kabir and the incinerated remains of the garden shed. When Jenny, Margaret, and Sebastian arrive, Kabir informs them of Michael's confession and his act of vengeance. Sebastian regrets that Michael escaped trial for his crimes and wishes he had been able to turn him in. Margaret, however, grabs a shovel and urges the others to help bury the charred body, reminding them that they all sheltered Michael. Jenny chooses to protect Kabir by burying the body and blaming the burned shed on vandals. She refuses to offer prayers over the grave; instead, Kabir reads the letter from the buried box, which addressed Michael as a trusted protector and begged help for the doomed congregation. A few days later, after others have dispersed, Ayesha arrives with good news about her exams. Only Margaret remains, waiting to depart. She gives Kabir's parting gift of a diamond to Ayesha and relays his message that she should use it to finance her dream of world travel.

The failure to prevent violence or articulate an alternative to revenge suggests that the traditional moral institutions of British society are at a loss to deal with the actuality or implications of violent conflicts in which the rule of law and moral inhibitions have broken down. As Gabrielle Griffin points out, the 'dying British institutions' of the Anglican Church and the remnants of the colonial empire are shown to be ineffective in 'protecting those whom they seek to govern' (229). All the characters have come to the church in their need for healing, forgiveness, and hope, but it proves inadequate. The kind-hearted but ineffectual Jenny represents an institution that has lost moral authority and increasingly loses ground to secular pursuits such as personal fitness. Jenny welcomes the refugees from the world's brutality to the churchyard sanctuary, cooks the occasional hot meal for them, and encourages them in ordinary ways, such as giving Sebastian the opportunity to teach a class. She fails, however, to demonstrate strength or articulate moral standards. She postpones telling Kabir when their appeal against closing the church is denied, and thus turns his disappointment into anger. The moral issues that enter her world when Michael's identity is revealed stun and silence Jenny. Her one moment of anger, in which she lashes out at Sebastian for apparently expecting to benefit professionally from turning Michael over to authorities, implies that she seizes on an issue of media exploitation as a distraction from the overwhelming consciousness of evil. She acquiesces in her grandmother's desire to bury the evidence of what has happened, offering no comment on Michael's life and death or the questions of complicity.

As all the adults are expelled from the refuge they have created and destroyed, the teenage Ayesha remains as its sole heir. Since her father is buried there, she continues to hold a stake in it. She, like the churchyard, is at a point of transition. Having just completed her exams, she wants to leave home and travel abroad. Memories of her friends from the churchyard – especially Kabir, who encouraged and even wanted to adopt her – may provide confidence and strength. The valuable but tainted diamond, which she has been given without being told of its origin, forms an ambiguous legacy. Ayesha has survived the death of her father and an unknowing brush with evil, though the institutions of home and church have largely failed her. Now this representative of the post-Thatcher generation will have to find her own way in the adult world. The play's generational divide conceals from Ayesha the horrific histories and ongoing suffering of her older friends, but make it clear that she is entering a world shaped by these histories and the suffering associated with them.

World Music by Steve Waters, produced by the Crucible Sheffield in 2003, focuses on a British member of the European Parliament who interprets the African genocide in the context of his own experience in Africa. A nation with the fictional name Irundi, and tribal groups similarly designated Muntu and Kanga, has experienced a recent conflict like that of Rwanda. Events included the slaughter of Kanga by the Muntu majority, the movement of exiled Kanga into the country, and the bloody overthrow of Muntu control. In the wake of these events, the British MEP, Geoff Fallon, seeks recognition and emergency aid for Irundi, but his colleagues in the British delegation are reluctant to commit to any action. Set primarily in present-day Brussels, with flashbacks to Irundi in 1980, action occurs in a fluid mélange of overlapping scenes juxtaposing personal and official meetings.

Geoff has devoted himself to African issues since his visit to Irundi in 1980 as a young man. Always wearing the same rumpled suit, he has long subordinated personal and family matters to a one-man crusade promoting African autonomy and development. When his 19-year-old son visits, Geoff has difficulty focusing on him, but cannot disengage his attention from their waitress, a young African woman who is revealed to have fled from Irundi and who accepts Geoff's invitation to stay with him. In the meetings with his colleagues, Geoff insists that the reports of mass killings have been manufactured. Viewing Africans solely in the context of exploitation by Western imperialists, he argues that tribal differences were used by colonial governments to maintain control, and do not cause hostility in present-day Irundi. He brings to Brussels his Irundian friend Jean Kiyabe, who gives a speech emphasizing Africa's exploitation by Europe. He personally sponsors Jean's request for asylum and raises funds for him to join family in Canada.

Geoff's expertise gives him a great deal of credibility within the British delegation, especially since none of the others are 'up to speed' on African nations that have 'too many names with too many consonants' (22). When Geoff tells the other delegates that he has seen Muntu and Kanga living side by side 'without one iota of evidence . . . to suggest the capacity for the kind of wholesale calculated violence that we are now being asked to accept' (22), he is sincere but unaware that his personal involvement has blinded him to reality. Flashback scenes show him as a young volunteer teacher in Irundi, staying at Jean's home, enjoying beer and manly conversation with the older man, and submitting to an impromptu ceremony mingling his blood with Jean's to signify brotherhood. Geoff's championing of Jean and his sheltering of the Irundian refugee Florence derive from the sense of obligation he developed as a result of these experiences. He pressures his delegation to support Jean's position, but they begin to have doubts. He becomes sexually involved with Florence, because she reminds him of Odette, a Kanga servant in Jean's house many years ago. Florence begins to nudge Geoff towards reality. When he tries to elicit stories of victimization, Florence shocks him by confessing that she was a killer who 'snatched babies', denounced classmates, and drew up 'lists of names of those to be exterminated' (72). Eventually he also learns that Jean engaged in mass killing, through documents given to him by colleagues. Geoff continues to deny Jean's involvement until he finds Odette's name on a lengthy list of victims.

Geoff's blindness to identity systems and power relations within a society he claims to know well offers a troubling assessment of the difficulty of understanding African societies from the standpoint of Western experience. On a very basic level, it emphasizes the dangers of attempting to comprehend the causes of violence and judge the perpetrators of it from the distance of a separate history and culture. The play's world music metaphor, in which African music 'is in tune with its own tunes but not with ours' (78), points to the necessity of becoming familiar with a different musical scale in order to interpret African music. It suggests that experience within a culture is necessary in order to understand the history and politics of a people. At the same time, the play cautions that adopting the perspective of people of a particular culture may interfere with objective evaluation of their situation. In its portrayal of Geoff, *World Music* suggests that his type of international activist, passionately advocating a cause but lacking a social and family life, may be particularly naïve about interpersonal dealings and inclined to meet personal emotional needs through aiding those perceived to be victims. The flashback scenes, viewed against the knowledge of the genocide, present strong hints of a campaign of ethnic violence already in its early stages – hints which Geoff, as a young newcomer enthralled with the novelty of the culture, could not interpret correctly. Interestingly, the play offers sympathetic portrayals of both Jean and Florence, thereby giving the audience an opportunity to be influenced by their appearance of innocence and victimization. *World Music* points to flaws in the British response to wars and political rivalries in Africa, but offers no suggestions for new patterns of interaction with other nations. It comes close to implying that the problems are insoluble and thus provides an international dimension to the post-Thatcher attitude of futility in relation to political involvement.

Though conflicts continue to emerge in Africa, the attack on New York's World Trade Center and the subsequent American-led invasions of Afghanistan and Iraq have captured the attention of media, politicians, and the public in the English-speaking world since 2001. As a result of the invasion of Afghanistan in 2001, and subsequent military operations in Afghanistan, Iraq, and other countries, American forces have captured hundreds of individuals suspected of involvement with Al-Qaeda or the Taliban. Early in 2002, American forces began moving detainees classified as 'illegal' combatants (to distinguish them from combatants with rights under the Geneva Convention) to a hastily constructed military prison on a naval base in Guantanamo Bay, Cuba. Inside Guantanamo more than 650 adult and juvenile detainees of various nationalities were subjected to interrogation and indefinite detention. Amid reports of mistreatment and torture, and with the United States refusing to give information on the number of detainees, their names, or accusations against them, Guantanamo was condemned by respected human rights organizations and hit with legal challenges. Though approximately 250 detainees have been released, the prison continues to hold the majority of its original prisoners, with no indication when they might be charged, tried, or released.

Guantanamo was commissioned by the Tricycle Theatre to make visible the situation of five detainees with British citizenship. Produced in 2004, this verbatim play was written by journalist Victoria Brittain and novelist Gillian Slovo, using material from interviews, letters, official documents, legal briefs, and public statements. Through interviews with British detainees who, after pressure from the British government, were released in early 2004, the play goes inside the secret and heavily fortified prison to give audiences a glimpse of the daily life inside. Detention and interrogation in Guantanamo subjects these men to dehumanizing conditions, separation from their families, boredom, mistreatment, and denial of access to information about the reasons for their detention. The detainees, their family members, and their legal representatives tell the men's stories and argue their innocence in this play, which is the only public forum to which they have had access.

The men represented in the play come from different backgrounds and experiences, but their interwoven narratives create a unified protest against the detentions. Moazzam Begg, the affluent and educated son of a banker, operated an Islamic bookshop in England and then went to Afghanistan with the dream of starting a school. Unable to get permission for his school from the Taliban, Moazzam began installing water pumps in villages. When the American bombing started, he fled with his wife and children to Pakistan, but was seized in the middle of the night by Pakistani and American soldiers and driven away in the trunk of their vehicle. His whereabouts were unknown until the Red Cross notified his father one month later that he was in the custody of the Americans. Wahab al-Rawi, an Iraqi émigré with British citizenship, was arrested by the Gambian secret service after he and his brother and two partners had gone to Gambia to start a peanut oil factory. Wahab was released, but his brother Bisher was sent first to a US prison in Afghanistan, and then to Guantanamo. Jamal al-Harith went to Pakistan in 2001 to study Islam. He was a passenger in a truck when the truck was commandeered by armed Afghanis, who took him to Afghanistan and turned him over to the Taliban. Suspected of being a British agent, he was interrogated by the Taliban. When the Americans arrived soon after and took over the prison, they promised to release him, but instead moved him to Kandahar and on to Guantanamo.

Letters referring to the time spent in detention indicate the way in which, to the increasingly desperate or despondent prisoners and their families, months of detention become years. Scenes juxtapose the detainees' letters with statements by government officials, human rights representatives, and a man whose sister died in the attack on the World Trade Center. American Secretary of Defense Donald Rumsfeld, in a press conference, describes the detainees as 'people who had been through training camps and had learned a whole host of skills as to how they could kill innocent people' (30). British Foreign Minister Jack Straw announces the February 2004 release of five Guantanamo detainees who are British nationals. Human rights advocates condemn detention without trial in Guantanamo and the British Belmarsh Prison, and point to broader implications of these violations of basic rights. Tom Clark, whose sister was killed in the terrorist attack, expresses outrage at the injustice done to innocent people in the name of combating terrorism.

The detainees cope with their situation as long as they can, but the play shows increasing censorship of Moazzam's letters, his father's anguish at the cessation of letters, and the report of human rights lawyer Gareth Peirce that Mozzam was believed to have been 'driven into mental illness' by prolonged solitary confinement (58).

Though this play deals with issues that can only be resolved at the highest levels of government, its strong character development aims at the individual consciousness of audience members. Its most important achievement is breaching the barrier separating Western audiences from the people imprisoned by their governments. Vivid portraits of the prisoners and members of their families create empathy. The men are present as sons, brothers, and fathers who worry about the well-being of loved ones, remind a parent to renew their motorcycle insurance, ask for contact lens supplies, agonize over the suffering brought on their families by their detention, and try to maintain some hope despite being reduced to powerlessness. These portraits arouse anger and questioning of the system that has detained these men for years, labelled them 'terrorists', but never charged them with specific crimes. Despite recent legal challenges in the United States, the system continues, and Bisher al-Rawi, the Iraqi citizen in the group, remains in the Guantanamo prison. The last of the detained British citizens, Moazzam Begg and Jamal al-Harith, were released in January 2005. The play's relevance is undiminished, as the weak or non-existent evidence of terrorist activity for these specific detainees points to questions about the entire population of these prisons. The play urges each individual in the audience to become informed about the situation and exert political pressure towards restoring human rights to the imprisoned.

Asylum seekers

Conflicts around the globe have produced a large number of refugees seeking political asylum in democratic nations such as Britain. According to a special BBC report published in 2003, one person in three hundred globally is a refugee. Applications for asylum have soared in the United Kingdom, from 4223 individuals in 1982 to 103,080 in 2002,[2] with the majority of applicants being young males from nations in conflict, such as Zimbabwe or Iraq ('Asylum Facts and Figures'). Although only one out of four applications is granted, the increased costs of immigration control, along with demands on the NHS, social services, and the state education system, have created resentment. The influx of refugees has also generated fears of terrorism and crime. The Blair government has responded to these concerns with measures that seem punitive to many. Asylum seekers without housing or resources are placed in detention centres while their claims are reviewed or appeals considered. Both the prison-like conditions at the detention centres and procedures for reviewing claims and appeals that do not take into consideration the circumstances of people forced to flee for their lives have been criticized by those to the left of New Labour and organizations concerned with the welfare of refugees.

Three recent plays address the plight of asylum seekers, focusing on both the violence from which they escaped and the difficulties of gaining residency and making a new life in the United Kingdom. The first of these to appear was *The Bogus Woman* by Kay Adshead, which was commissioned by Red Room, premiered at the Traverse Theatre in 2000, and performed at the Bush Theatre in early 2001. During the same period in 2001, the Royal Court Theatre opened *Credible Witness* by Timberlake Wertenbaker. Sonja Linden's *I Have Before Me a Remarkable Document Given to Me by a Young Lady from Rwanda*, produced by a new company called Iceandfire, opened at the Burton-Taylor Theatre in Oxford in 2002, went on to a four-week run at London's Finborough Theatre in 2003, and toured the United Kingdom in 2004. These three plays show commonalities, each of them written by a woman and featuring a female central character. Elaine Aston's argument that the plays by Adshead and Wertenbaker exemplify a feminism that 'situate[s] the narrative of asylum within a matrix of gender, race, and nation' and urge audiences to 'shift their view of an imperialist past and learn to see themselves as inside (not outside) asylum histories' (6) applies as well to Linden's play.

The three plays encompass only two different regional conflicts, but use quite different styles. *The Bogus Woman* employs poetic language and a non-realistic style to present the experiences of a young African woman as she goes through the process of applying for political asylum in Britain. As a one-woman play, this piece filters the viewpoints of the entire cast of characters through the narration and performance of the actor whose primary role is that of the young woman. This choice intensifies the focus on the main character and raises the stakes in her struggle. In scenes that follow her through Campsfield and Tinsley House detention centres, as well as a brief period of relative freedom when a volunteer finds her a room, the young woman meets sympathetic people without power and powerful people who have no sympathy. Despite the atrocities she suffered in retaliation for articles she had written as a part-time journalist, she demonstrates a strength and will to live that are rooted in her desire to communicate the story of family members who are now dead. Her unwavering determination contrasts with her continual move-ment, from detention centre to hospital, to another detention centre, to a rented room, and finally to the street. Her struggle for survival as an unwanted refugee takes a physical toll even greater than that she suffered in her own country, as it starves, exhausts, ages, and cripples her. She loses her only possessions – photographs of her family and the shawl in which she had wrapped her newborn daughter. At the end of the play, having been refused asylum, deported, and killed with those who shel-tered her, she has ceased to exist except as a warning of the tragic fate that may await refugees sent back to the countries from which they escaped.

Wertenbaker's *Credible Witness* employs a more complex plot and frames its refugee characters within a lengthy history of persecution rather than a recent genocidal conflict. The prologue scene, set in northern Greece, shows Alexander Karagy, a young Macedonian history teacher, attacked by shadowy figures later understood to be local police. Subsequent scenes, presented realistically and chronologically, show Alexander living illegally in England and his mother Petra arriving in England seeking him after a three-year absence and no communication. Alexander at first works with refugee children, teaching them to reclaim their personal histories, cry for what they have lost, and plan for the future. Fear of arrest and deportation gradually reduces Alexander to menial jobs and living rough, but he still maintains contact with two of the children and encourages them to build new lives. Petra claims asylum and finds herself in a detention centre, where she develops a family-like closeness with other refugees, learning about the horrifying tortures some of them have endured and encouraging their hopes for a better future. She goes on a hunger strike, which publicizes the conditions of the detainees and brings her son to her side. Because he has abandoned his Macedonian heritage – 'lost his his-tory' (2002: 22) – Petra bitterly rejects him. By the end of the play, however, she comes to see his and her own identity in new terms.

The central character in *I Have Before Me a Remarkable Document Given to Me by a Young Lady from Rwanda* focuses on the artistic and personal relationship between Juliette, a young woman who has been granted political asylum in Britain, and Simon, a writing tutor at a refugee centre. Juliette and Simon share the action in dialogic scenes and soliloquies, as each reveals the changes in perspective wrought by the other. Juliette attempts to come to terms with her past while struggling to adapt to life in a new country. Simon, troubled by personal and career failures, seeks meaning in helping Juliette and finds himself attracted to her in ways that complicate their relationship. She wants to write a memoir that tells the truth of what happened to her family, but finds it extremely difficult to revisit the terrible scenes of murder and rape, even in memory. When Juliette discovers that a younger brother is still alive, living in Uganda, she applies for asylum on his behalf, but he is refused entry. With the help of Simon and his wife, she goes to Uganda, assists her brother in finding a safe home and educational opportunity there, and returns to Britain with a renewed will to live and tell the story of her family and nation.

These three plays present notably similar portraits of refugees seeking asylum. All are, in Wertenbaker's words, 'possessed by history' (2001: 23). The young African woman in Adshead's piece finds strength to live in her desire someday 'to write a history... of my people' (45). Petra has instilled Macedonian history into her son and encouraged separatist resistance, but by the end of the play she recognizes that even her own history is not separate from those of other nations. She tells the director of the detention centre that she has been mistaken in a 'walled' concept of history and urges him to look around at the refugees from Somalia, Sri Lanka, and Algeria and understand, 'we are your history now' (2001: 63). Juliette in Linden's play is writing a 'remarkable document' that will testify to both the pride and the tragedy of Rwanda (19).

Each of the refugees feels that her or his life exemplifies a particular history. Each feels that surviving conflicts in which whole families and communities were killed has created both a unique perspective and a unique responsibility to record and preserve history. Having been propelled out of familiar cultures into a strange world in which their identities are submerged in the blanket term 'asylum seeker', all seek to revive and make visible their identity and the culture in which it was formed. At the same time, all inhabit new situations that deny their previously established identities. Their experiences in the immigration system bring out new aspects of personality and develop new dimensions of identity. *The Bogus Woman* shows a tragic progression as the racism and insensitivity of the bureaucracy damage the young woman, while the desperate hope of fellow detainees and the well-meaning help of individuals prove inadequate to save her. Juliette in *I Have Before Me*... moves to a triumphant sense of integration, using her newfound confidence and fluency in English to express the treasured vision of her homeland as 'the country of a thousand hills' (64). Petra and Alexander in *Credible Witness* learn a new, non-nationalistic form of identity. The dying Petra will not be able to move into the new paradigm, but she sees that the rigid patterns of the past have lost their meaning and gives her son blessing to live his life in accord with a changed understanding of identity.

In their negotiations of identity within a system that deprives them of personal power, the refugee characters demonstrate unusual strength rooted in a sense of personal and collective pride. As a result of political violence, they have lost loved ones, homes, and even articles associated with home. Their non-citizen status has deprived them of important aspects of selfhood. Nevertheless, they maintain their dignity and assert their human value. When Simon first comprehends the extent to which Juliette's activities are limited by lack of money, his unconcealed pity makes her want to reassert pride. She soliloquizes, 'We Tutsis are proud, we don't want nobody to know if we have problems' (32), and describes how a family without food would place a cooking pot on the stove to create the appearance of cooking. Petra considers her son a descendant of Alexander the Great and, remembering her Bulgarian great-grandmother who resisted Greek power, describes herself as one of 'the women who refuse to kneel' (2001: 27). Strictly confined after a protest, and monitored in the toilet and shower, the young woman in *The Bogus Woman* maintains her insistence on human rights, writing complaints and demanding to see her solicitor. She, like Petra, also stages a hunger strike, putting dignity above life itself.

The immigration system constantly undermines the strength, dignity, and sanity of the asylum seekers. Petra insists 'I do not want an asylum' (2001: 11), thinking she is being sent to a mental hospital. The detention centre has a medical aspect, but medicines are used to control rather than heal, and the confinement actually creates illness. Denial of their humanity involves the detainees in resistance, but the young woman from Africa discovers that protest brings even harsher control and humiliation. When she is monitored while bathing, her intense shame provokes her to violently attack her own body. Juliette, though not confined, nearly loses her will to live when her brother is denied entry into Britain. The bureaucratic system based on colonial models of control disdains the languages and cultures of the people brought into it. The young woman's case suffers from poor translation that hinders identification of the newspaper that employed her, and after the protest her personal file unaccountably disappears. To an official who ignores her assertion of nationality, Petra complains, 'you try to disappear my country' (2001: 12). Simon urges Juliette to write about her family but is oblivious to the pain involved in remembering them. All three plays depict an arbitrary and opaque system that turns a blind eye to the terrifying dangers in the asylum seekers' homelands.

The plays manifest their political position through appeals to individual conscience. They generate empathy through the humanist construction of admirable characters who have survived extraordinary violence and now encounter institutional callousness and incompetence in Britain.

They suggest that individuals can counter the failures of the immigration system by showing the help provided by those who relate to the refugees with sympathy and acceptance. At the same time, they show that these individuals' lack of power limits their impact. Adshead's young woman is befriended by a volunteer who finds temporary accommodation for her so that she can leave the detention centre; unfortunately, this volunteer does not own the house and cannot prevent the young woman's eviction when its owner decides to house Kosovan refugees there. The young woman's solicitor continues working on her case, even when money runs out, but fails to establish the validity of her claim. In *Credible Witness*, a detention centre guard, while resolving wryly to 'become tough' (2001: 53) next year, brings in food for a festive New Year's Eve celebration with the refugees, but notes that kindness 'doesn't help you get on' (2001: 54). Though Simon develops a love for his protégé that threatens to become exploitative, he provides psychological support and tangible aid to Juliette as she writes her life-restoring memoir. Simon's wife solves Juliette's most critical problem by finding a way to finance Juliette's trip to Uganda to arrange for her brother's care; her generosity simultaneously helps Simon overcome personal difficulties and keep his relationship with Juliette within ethical bounds. The impulse to generosity can occur even in those who oppose immigration, as Paul notes when he tells the refugees that much of the food for the New Year's feast was donated by shopkeepers who have been known to 'threaten to beat you up' (*ibid.*). These few individuals, however, lack power to change the system or create alternatives for refugees. The plays therefore seem to aim at intensifying individual opposition to the current system for dealing with asylum seekers, to the point that it will galvanize change. The Committee to Defend Asylum Seekers, which started in 2001, provides organizational support for refugees in Britain. Meanwhile, with fresh waves of refugees arriving in Britain and other countries, the treatment of asylum seekers has become an international human rights issue.

Terrorism

Whereas terrorism in the Britain of the 1970s and 1980s generally arose from the conflict in Northern Ireland, terrorism in post-Thatcher Britain most often originates in conflicts outside Britain. British citizens have been both targeted by and randomly caught in the violence related to conflicts in the Middle East and Africa. *Talking to Terrorists*, compiled by Robin Soans, produced by Out of Joint, and first presented at the Royal Court in April 2005, attempts to look at terrorism from both sides – perpetrator and victim. This verbatim play uses material from interviews conducted by actors who subsequently played roles created from this material. Those interviewed include citizens from both sides of the conflict in Northern Ireland and representatives of ethnic and other political divisions in Iraq, Palestine, and Uganda, as well as members of the diplomatic corps and survivors of terrorist acts. All of the characters are based on actual people. None of the interviewees is identified by full name, though it is possible to guess at some identities, especially those of the Archbishop's Envoy who was kidnapped in Lebanon and the bomber and survivors of the 1984 attack on the Brighton Grand Hotel during a Conservative Party conference. Each actor plays several roles.

In its attempt to stage dialogue between people unavailable to talk with one another, the play alternates monologues to create a sense of communication where none actually exists. Perhaps inevitably, both sides address the audience more than each other. Accounts by terrorists from around the world emphasize their status as victims. Rebel forces in the Liberian civil war, for example, kidnapped children and used them as foot soldiers in their conflict. They forced the children to kill their parents and indoctrinated them with the belief that they were bullet-proof. A former child soldier from Uganda describes playing in the road to decoy an army vehicle into an ambush and participating in a mass murder, adding 'we killed with total commitment . . . to please our bosses' (57). This young woman also experienced forced sex with her commander, and comments bitterly on their supposed fight for freedom, 'as if there were any freedom to fight for' (58). An IRA member describes the impression made on him by violence in Belfast when he was 13. A psychologist speaks of adolescents' vulnerability to the appeal of outlawed movements. A Kurdish resistance fighter describes the injustices that led to his involvement and the torture he suffered after being caught.

The foregrounding of children forced or enticed into a life of violence serves to soften the line between terrorists and victims, shifting the responsibility for violence to an unseen oppressor or a self-perpetuating cycle. The play does not explore differences between the justice claims of a leader in the Palestinian Al Aqsa Martyrs Brigade, who was wounded and locked inside Bethlehem's Church of the Nativity for 39 days, and the Archbishop's Envoy, who was kidnapped while on a mission of mercy and held in solitary confinement for 'one thousand seven hundred and sixty-three days' (52). It does not confront the implications of the wide gap between the experiences of perpetrators and victims in the 1984 Brighton bombing. The IRA terrorist who planted the bomb offers this statement: 'Of course I regret the suffering I caused but circumstances made our actions inevitable' (85). His detailed recollections demonstrate failure to empathize with the victims, consider alternatives to violence, or accept personal responsibility. The detachment evident when he says, 'Sitting here now . . . it's like talking about someone else' (86) contrasts palpably with the experience of the victims. They still live very close to the event, speaking of the ways in which it currently affects their actions and referring to friends who died or were injured. Unforgettably, the former cabinet minister whose wife was permanently disabled in the bombing speaks of sleeping with a hunting gun under his bed in an attempt to combat his persistent sense of vulnerability.

The transformations taking place continually, as actors shift from one role to another, provide the power dynamic that is both hopeful and chilling. On one hand, they point to the kind of change exemplified in the Kurdish militant and the former girl soldier from Africa. Both of these individuals have survived the violence of their past, found refuge, and embraced new values in Western nations. On the other hand, a group of young British Muslims talk about the attention paid to extremists in their group: 'Of the Muslim community in Luton they're not even one percent . . . the fanatics . . . but they're listened to more than the rest of us put together' (34). The potential power in such attention suggests that those with even small grievances might turn to terrorism. The final scene, of a schoolgirl in Bethlehem whose feelings for the victims of the Twin Towers' destruction change from 'sorry' to 'happy' after one of her friends is killed by an Israeli sniper, foreshadows the continuation of terrorism in a world damaged by conflict and oppression. *Talking to Terrorists* takes the position that only by talking with the perpetrators of the violence can such violence be ended. The variety of perspectives and combined experience in these performed interviews provide a brief lesson in how terrorism develops but do not pose the most difficult questions. The artificially constructed dialogues remove the responsibility for resolving the conflicts from the terrorists or their victims and place it on the individual audience member. The haphazard and essentially powerless nature of this caught-in-the-middle position became clear in the real events of July 2005 when terrorist bombings of the London underground trains and London Transport bus occurred even as the play was in the midst of its second run at the Royal Court Theatre.

Global business

The globalization of commerce and industry gained widespread public attention during the Thatcher period. Whether it resulted from deliberate policies or was an irresistible development, as Margaret Thatcher suggested in the TINA (there is no alternative) doctrine, the global economy may prove to be the most enduring legacy of the period. Globalization promotes not only the flow of goods, but also that of capital, people, and intangibles like specialized knowledge, across national borders. Aspects of globalization, such as the power of multi-national corporations, the increasing wealth gap between rich and poor countries and individuals, the exploitation of developing nations, the displacement of indigenous cultures by consumer culture, and the move-ment of jobs from high-wage to low-wage regions, have created political opposition. Recent plays about globalization highlight the huge power disparities between multinational businesses and the individuals affected by them. These plays often combine or contradict empathy with sarcastic or absurdist humour that points to the impossibility of resisting the power of the multinationals. All highlight the waste of lives and destruction of families and communities by associating global busi-ness with death.

Winsome Pinnock's *Mules*, commissioned by Clean Break Theatre Company and performed at the Royal Court Theatre Upstairs in 1996, uses the illegal but enormously profitable global business of drug smug-gling to illustrate the power of multinational enterprise. An all-female cast highlights the exploitation of women in global business. The Jamaican woman Bridie has moved from rags to riches by recruiting and overseeing 'mules', young women who transport illegal drugs by hiding them inside body cavities, under wigs, or taped closely to their bodies. This work exposes the young women to the risks of arrest and imprisonment, addiction and even death if the condoms containing the drugs break and release a toxic concentration into the woman's system. Consequently, Bridie seeks her mules among the truly desperate – poor women in developing countries and teenage runaways in large cities. These women find in the work not only an alternative to begging, but also a sense of accomplishment when they complete the dangerous trips and a taste of power and privilege when they are encouraged to indulge in high-priced clothing and accessories.

That desperate young women become mules and suffer the all but inevitable consequences does not come as a surprise. As one says of Bridie, 'she was the only person I ever meet in me life that offer me hope' (71). The play's unexpected development is the young woman who, after an initial run, refuses further work. Lyla, a Jamaican woman recruited along with her sister after they fail at selling underwear in a market stall, does not trust Bridie but goes along with her sister. When her sister is arrested, she figures out that Bridie betrayed her as part of an arrangement with authorities that allows most of the mules to get through. Lyla returns to her town in Jamaica, takes care of children, and earns what little money she can locally. When Bridie visits and brings food, including a tin of corned beef, Lyla maintains her independence, insisting that she has plenty of fresh food and considers the corned beef fit only for dogs. Though she does, out of necessity, accept the food and a job on a nearby marijuana-growing plantation, Lyla remains close to home, teaches children to 'play gentle' (58), and shuns the consumer culture that traps women into working as mules.

Though *Mules* deals with an illicit business, it clearly implies that drug smuggling is fundamentally just another business. When first seen at work finding someone to handle an assignment in Amsterdam, Bridie could be a manager in any sort of corporation. When, however, she introduces her business to the audiences, Bridie uses the imagery of death. She speaks of being haunted by the ghost of 'some poor dead girl' (5), but says she can deal with it. She demonstrates her business competence in the matter-of-fact reaction to the death of one mule and the mock execution of another who absconded with drugs. Though Bridie seems to be in control, she is found, towards the end of the play, badly beaten. An employee who had regarded her as a role model realizes that Bridie too is 'nothing but a mule' (67). By contrast, the work Lyla chooses is full of life. She labours in the sun, cares for children, becomes a mother herself, and celebrates with her sister when she returns home after release from prison. The two sisters, at the end, live in poverty but enjoy security and love within their family and community. They provide an example of individual resistance through separation from global profit-making enterprises that threaten their autonomy and even their lives. That poor and uneducated women generate such resistance fosters hope that ordinary individuals may protect themselves from these threats. This message of hope stands out uniquely among post-Thatcher plays about global business.

Gagarin Way by Gregory Burke, which was a co-production of the Traverse Theatre in Edinburgh and the National Theatre in 2001, shows the global corporation under attack by a pair of Scottish factory workers. Eddie and Gary, who discuss the writings of Sartre and quote from *The Revolutionary Catechism*, are angry about losing control of their lives to outsiders. They plan to kidnap a representative of the Japanese corporation that has taken over their factory, kill him, and leave a written manifesto on the body. To secure the cooperation of Tom, a recent university graduate working at the factory as a security guard, Eddie tells him they are transporting stolen computer chips for sale on the black market. When Tom opens the gate for Gary's van, Eddie sends Tom on his way while Gary unloads their unconscious hostage. Tom, however, returns for a forgotten hat, and he too becomes a hostage. These four employees of the corporation engage in an improbable conversation about politics, economics, and history. The lively dialogue, which the locals deliver in the Dunfermline dialect, brims with wit and irony. Eddie says he considered the wide range of choices open to him and concluded, 'I've always been interested in violence' (41). Finding that the pleasure of 'recreational violence' has diminished, he decided to 'try something more idealistic' to see if the combination of politics and violence 'still had a contribution tay make tay the modern workplace' (*ibid.*). Gary has political motives. He reveals they had considered kidnapping a politician, but decided there was 'nay fucking point' in that, because 'politicians don't matter a fuck anymore', whereas corporate businessmen 'hold absolute power in the world' (64). The two argue continually over whether the killing will communicate a message of protest or test the climate for political violence. The two hostages show different attitudes towards survival. Tom nervously attempts to steer conversation away from violence, enquiring of Frank, as if they were making small talk at a party, 'Do you... think that devolution's been a success?' (67). Frank, the corporate manager, maintains a world-weary calm, warning the gun-waving Eddie that 'hitting the target' is difficult (65).

In the course of a conversation that skirts the limits of comic absurdity without losing its suspenseful focus, the two would-be terrorists learn that they have failed to hit their target. They had intended to take a Japanese manager as their hostage, but he turns out to be from nearby Leven and further informs them that an American corporation recently acquired the factory. Frank agrees with his captors on most points, denouncing corporate rhetoric about being 'one big happy family' (58) and observing that 'capital comes, capital goes' (61). All four men actually share a working-class history, with fathers and grandfathers who were miners, and a familiarity with the area's radical history, which has left some streets, like Gagarin Way, named for Soviet heroes.

Within this surreal context, oppositions clash and merge. Each fact Gary and Eddie learn about Frank places him closer to their own experience, rather than in a separate and alien sphere. Gary wants Frank to defend his work, the company, and the economic system of which they are a part, because he needs something definite to oppose. Eddie just wants to kill, scorning the concept of meaning. Frank patiently and knowledgeably discusses anarchism and other forms of political protest, emphasizing the impossibility of meaningful protest against the overwhelming power of global corporations. Tom naively tries to help by suggesting that one need not choose between socialism and capitalism, because 'it can be both' (85). Gary has planned to avoid shedding blood, as becomes clear when Eddie fires the gun and discovers it has no bullets. Eddie, however, produces a knife, and he and Gary fight for control of it. Eddie gains control and kills both hostages – the manager and the security guard meeting the same bloody end. The bizarre oscillation between absurdity and meaning, along with all the oppositions that have been introduced – rational argument versus brute force, local versus global forms of power, history versus the present, political activism versus apathy – dissolves in a senseless double murder. This play, in common with *Mules*, pits ordinary people against the power of multinational business, but in this case the result is nihilistic violence. The outcome, as in *Mules*, associates global business with death; however, in *Gagarin Way*, all potential for life is extinguished. As the two collaborators face the aftermath of their crime, Gary stammers a weak denial of his power to act, while Eddie makes it clear that he will kill Gary if he cannot 'rely on' him (91). Thus, their attempt to act has only further alienated and isolated each of them, while having no effect on the corporation they originally targeted. If not entirely dismissing the potential for political action, this play makes an extremely bleak assessment of the ability of individuals, acting alone or in concert, to make even a symbolic protest against the oppressive power of global corporations.

Wild East by April de Angelis, produced in 2005 by the Royal Court Theatre, places its critique of global business in the context of a job interview. The interview seems initially to highlight the opportunities available in multinational corporations. The play, however, uses absurdism to heighten the uncertainty and disconnectedness inherent in the interview situation. Frank, a recent anthropology graduate hoping to land a job that would locate him in Russia, where he has a girl friend, arrives for the interview that has been scheduled for early evening in an ultramodern office in Surrey. Frank makes his first mistake when he interacts with a woman he meets in the waiting area as if she were a fellow interviewee; she turns out to be one of his interviewers. The interviewers introduce themselves with their formal titles and ask standard questions, such as 'what made you turn to anthropology' (12) or 'tell us about a situation where you were presented with a problem you had no idea how to solve, and how you eventually solved it' (17). Frank responds with long-winded stories revealing his aimlessness, lack of social skills, and poor self-control. For example, he reveals that he studied anthropology because he initially confused it with archaeology, and confesses that he was suspected of stealing a priceless artefact while working on an archaeological dig. Dr Pitt and Dr Gray, the two women conducting the interview, exhibit increasingly strange behaviour, and the role-playing exercises they lead disintegrate into emotional brawls. Frank eventually understands that the two women are so isolated and powerless that they see everyone, including him and each other, as allies or enemies in the continual battle to keep their jobs.

In the course of the parodic high-pressure interview, Frank reveals and then destroys his secret identity. After requesting that the videotape be stopped, he produces from his pocket a small stone effigy of a bird – the artefact he was suspected of stealing. He reports that he was moved by the beauty of the effigy, but says he took it because it 'connected me back to something' and summoned him to the 'spontaneous vocation of the shaman' (63). Bringing out a marijuana joint, he invites the interviewers on a spiritual journey. When that experiment ends in a fight and Frank is informed that he is not suitable, he suddenly and decisively smashes the bird effigy.[3] Seeing Frank's willingness to sacrifice his own soul, as well as the culture represented by the artefact, Dr Gray ushers Frank into the corporation. They leave behind the distracted Dr Pitt, who vainly attempts to pick up the shattered pieces of the bird.

The metaphorical death at the climax of this play evokes a destructiveness that goes beyond the lives of individuals. The play views global business as a monster that dehumanizes its employees, exposes them to harm, and subjects them to soul-destroying competition. It tramples on the natural environment and shatters indigenous culture. This monster does all these things in pursuit of absurd goals, such as creating a Russian market for Russian-style yoghurt not manufactured in Russia. Individuals like the nerdy Frank, the emotional Dr Pitt, or the brusque Dr Gray lose their individuality when they enter the service of the corporation. They become disposable parts in a giant machine, to be replaced rather than repaired when they need maintenance or cease functioning at their peak level. *Wild East* goes for humour rather than emotion, but its humour is bitter, and the absurdity has an edge of desperate fear. This play begins with an individual who wants to join and profit from rather than oppose global business, but the future awaiting him builds on his utter surrender of his life's meaning and promises only that he will be used and discarded.

Britons abroad

The global economy has brought more frequent and casual travel across national boundaries, and with it new patterns of political interactions with other nationals. Three recent plays centre on the entanglement of British citizens in the politics of other nations, humorously contrasting the British character's identity and expectations of the world with perceptions of them in the host society and the actual conditions of the host nation. *US and Them* by Tamsin Oglesby, produced by the Hampstead Theatre in 2003, presents the special relationship between the British and the Americans in a new light. *The God Botherers* by Richard Bean, produced by the Bush Theatre in 2003, shows British NGO workers in sub-Saharan Africa promoting change at the same time that they are being changed by life in the host society. *The UN Inspector* by David Farr, produced by the National Theatre in 2005, is an adaptation of Gogol's *The Government Inspector* in which a would-be entrepreneur from England ventures into an unexpected situation in the capital city of a former Soviet Republic.

US and Them uses a personal friendship between a British couple and an American one as a metaphor for the relationship between the two nations. Martin and Charlotte have come to New York to see the city and seek a backer for Martin's invention, a robotic lawn mower built to resemble a sheep. Lori and Ed meet them by chance in a restaurant when Charlotte's coat is singed by a space heater, and Ed steps in to take charge of making sure the embarrassed visitors are compensated for the loss. Charlotte and Martin find the American couple 'friendly... so open... forceful' (22), while Lori considers the British couple 'charming' and Ed notes 'the pleasant surprise... that they are, in fact, warm-blooded animals' (23). After discovering that the two women share a common maiden name, the Americans invite the English couple back to their apartment for a nightcap. As the friendship develops, each couple spends time in the other's culture. The British are impressed and slightly intimidated by the luxurious lifestyle of the Americans, but Martin is gratified when Ed offers to back his invention with a company called Milcam. When the Americans visit England, Lori tours graveyards looking for ancestors, believing that Charlotte is a distant relative because of her maiden name. The couples struggle through disastrous meals on one another's home turf and uncomfortable conversations about American power and accountability in world affairs. Meanwhile, their children, 19-year-old Izzie and 20-one-year-old Jay, get acquainted.

The special relationship ends when the pet projects that had connected the couples come to unsatisfying conclusions. Milcam brings out a product that is remarkably similar to Martin's invention, but without acknowledging or paying him. Lori learns that Charlotte was adopted, and therefore is not genetically related to her. Though both couples experience disappointment, it is Ed who, with his characteristic direct-ness, breaks off the friendship. In a scene that begins the play and is repeated at the end, he simply announces at the end of an evening together, 'I'm afraid we have come to the conclusion, albeit with regret, that we don't want to be your friends anymore' (127). This statement opens a floodgate of resentment and blame, with each couple dispara-ging the other in terms that echo typical criticisms of their culture. Martin accuses the Americans of thinking they can 'go around *buying* everything' (130) and of behaving like children 'who think the world centres around you' (133). Ed sneeringly advises Martin and Charlotte to 'go back to your whingeing little country with your contempt and your sniping and your misery and your failure' (131). Martin retaliates by identifying Ed and all Americans with 'a trigger-happy ex-alcoholic who can barely speak English' (135). Ed punches Martin, but Charlotte, unexpectedly breaking into laughter after a bout of vomiting, has the last word, demanding, 'Who the hell do you think you are?' (137).

Though this friendship ends in disaster, the play's final scene indicates that neither the Americans nor the British will ever escape their close relationship. Izzie and Jay have bonded over a common appreciation of techno music and rebellion against parental authority. At the plays end, they are together in Izzie's flat preparing to phone Jay's parents with the news that Izzie is pregnant with his child and that they plan to marry. As the two young people argue lightly over whether to find out the baby's sex, and the phone rings across the ocean, the question that hangs in the air is, how a special relationship can be expressed when the parties to it have thought and said the worst of each other?

Richard Bean's *The God Botherers* presents its view of British-run non-governmental organizations in developing countries through two contrasting characters. The middle-aged Keith has been running the organization for 3 years, and has been abroad for 20 years as a foreign aid worker in various parts of the developing world. The work has become his identity, as he demonstrates by starting many sentences with reference to a previous placement, such as 'when I was in Bangla Desh in '95' (53). He recalls with pleasure a jungle assignment where toileting required a shovel 'not to dig a hole, but to kill the snakes' (16). Child support payments constitute his only contact with ex-wives and children in England. Laura is a 23-year-old newcomer, elated at her freedom and eager to share her experiences with a friend back in England. Inexperienced and impressionable, Laura voices naïve opinions formed from reading *Cosmo* and the 'Idiot's Guide to Islam' (29). Through scenes punctuated by monologues, Keith and Laura collaborate in surprisingly effective ways. Keith works through the authorities and eventually succeeds in getting a water tap installed in the village. Laura works locally, forming personal relationships. By dating the regional cell phone company's representative, she acquires cell phones for the village women and a computer for the NGO headquarters. Laura even manages to attract tourists to the village's annual festival celebrating female circumcision.

The two Africans who work closely with the NGO, Ibrahima and Monday, exemplify the gender oppression and religious divisions in their society. Ibrahima's husband throws kerosene on her and lights it after she attempts to vote, and threatens to kill her if she gives birth to another girl. Monday claims to be a witch doctor in the Poro sect, observes many Muslim customs, but wears a hat that identifies him as a Christian when he goes drinking. Monday is caught and flogged when he forgets his hat and drinks alcohol without wearing it. Angered by this experience, he returns to paganism and rebels against Islamic authority. The two Africans also illustrate the impact of the changes brought to the village by the NGO. Monday, who has earned money by ferrying water to the village on his bicycle, loses a source of income when the tap is installed. When the computer arrives, Ibrahima becomes the first villager with an email account and immediately receives an advertisement for penis enlargement. Laura, meanwhile, exposes her own superstitious beliefs by organizing an email prayer chain for Ibrahima's expected child to be male.

Despite their differences, both aid workers become one with the place to which they have been assigned. Their common danger, as conflicts escalate and Ibrahima's pregnancy advances, draws Keith and Laura into an increasingly close relationship with the Africans. Keith's involvement climaxes on Christmas day when he sets up a lighted tree and single-handedly confronts the local group of thugs known as the Diesel Boys, who have outlawed Christmas celebrations. His heroism allows Monday, who has cut off the beard of the local imam, to escape. Laura's involvement becomes clear in the final scene when she welcomes a new aid worker and reveals that she now runs the organization and has become Monday's wife. Like Keith before her, Laura has found meaning in a personal attempt to improve conditions in a developing country and has been remade by a society with extreme needs, ancient customs, and contemporary conflicts.

The God Botherers shares elements with Helen Edmundson's *Mother Teresa Is Dead*, showing similar outcomes for the would-be activists. Edmundson's play reveals that the British woman who tried to sacrifice her own identity to help the poor in India was confronted by fierce pride and determined strength instead of powerlessness. Whether she goes home or stays in India, Jane is fundamentally changed by this encounter. Laura adapts more easily to the world of the African village to which she is sent, but is no less altered. The culture Laura encounters has been little influenced by Britain, and her outsider status provokes curiosity rather than rejection. As she chooses to work with the village's culture rather than against it, Laura becomes one of the community, accepting in the process an identity and practices, such as eating bush meat, which she previously rejected. Both plays emphasize the power of foreign cultures to act on, transform, and ultimately humble the prideful Westerner and Western efforts to enact political and social change within their societies.

Though each of these plays about Britons abroad employs dark humour, the most farcical of the three is *The UN Inspector*. The action begins when the president of the ex-Soviet Republic tells his closest ministers of an email from an American contact warning of an imminent visit by 'a top UN official . . . with instructions to investigate rumours of government corruption, electoral fraud, and humanitarian abuses' (5). He thinks the visitor might already be in their midst, working under an assumed name. In a panicked review of their institutions, the president and his cabinet acknowledge serious problems, including no schools in the north, burgeoning slums around the capital city, and a hospital that has been converted into the set for a television programme. The government has privatized utilities, as required by the terms of its IMF loans, but the president has sold the water board to one relative, the gas board to another – who stripped its assets and sold it to a Texas conglomerate – and gave the electricity grid to his daughter on her birthday. The president has imprisoned a journalist who was investigating government corruption. Aides Robchinski and Dobchinski bring the news that an Englishman named Martin Gammon is staying at the new Marriott Hotel, and everyone concludes that he must be the UN official.

Martin Gammon, as we learn from his disgruntled sidekick, Sammy, is a former estate agent from England who hoped to profit from a property boom in the ex-Soviet Union. The six-month venture has failed, and now the two are out of money. They have reached their credit limit at the Marriott, where they are staying, and do not know where their next meal is coming from. When the president suddenly appears at the hotel, accompanied by several officers, Martin assumes they will be thrown in prison. When, however, the president gives him an envelope of money and invites him to stay in the presidential palace, he remarks, 'Well, this is more like it' (38). While the president pretends to believe that Martin is an ordinary businessman, Martin himself accepts the VIP treatment as no more than he deserves. Under the influence of plum brandy, he grows loquacious and brags of being a 'legend', a 'mover/shaker' (51), a columnist for three different London newspapers, and a literary wonder living in 'the biggest house in Hampstead' (52). He even claims grandly to be in charge of 'the whole Clapham branch of Foxton's', the estate agency from which he was fired (53).

Martin finds himself at the centre of frenzied attempts by all those in power to bribe and influence him. Each minister offers cash and confidential information about who should be held accountable for their country's problems. Though he begins to suspect that he is being mistaken for someone else, Martin holds on to the illusion that his 'real worth' is at last being recognized and that the bribes are 'some long overdue tokens of appreciation' (77). When Sammy urges departure, he reluctantly sends him to arrange a helicopter. Before they can go, however, a mob of small businessmen breaks through security, presenting grievances against the president and showering Martin with small gifts. In a more serious vein, the mother and a friend of the imprisoned journalist force their way in, plead for help in finding her, and describe the terrible conditions in their country. Martin sympathizes but is distracted by the flirtatious behaviour of both the wife and the daughter of the president. When he departs, he insists that the daughter, Maria, accompany him.

In the end, Martin's own action reveals the mistake. Following his departure, when protestors are being shot and the president is accepting congratulations on his daughter's engagement, the Head of Intelligence brings the bad news. His routine monitoring of email has picked up a message from Martin to a friend in London, detailing the entire mix-up, complete with insulting descriptions of all those present, for the friend to write up for a magazine. Furious at the thought that his humiliation will be published abroad, the president decides his only recourse is to prevent the helicopter reaching its destination. He orders it destroyed, under cover of a storm, even though it means the death of his daughter. As congratulations change to condolences, the Head of Intelligence has one more message to deliver: the UN inspector has arrived.

Despite their humour, the plays about Britons abroad, whether in the context of global business, aid work, or personal travel, contain sobering themes. They employ a changed concept of the world in which Britain is no longer insular. While recognizing that other nations have become more a part of British everyday life than in the past, they do not present a sanguine view of British power in international politics. These plays expand the stage on which the British nation, typically exemplified by a single individual, must act, but the vastness and complexity of global scenarios renders ineffective and even absurd any action the character might take. The precarious position of this lone individual is heightened by vivid conceptualizations of the dangers inherent in foreign settings. Foreign cultures and the conflicts that arise within them are opaque, difficult to penetrate, and liable to trap the one who does penetrate them. Only the environment of home offers the transparent familiarity that makes safety and comfort possible, but the possibility of maintaining home as an island of separateness and homogeneous culture has been destroyed by globalization.

In contrast to the plays about specifically national issues, those dealing with Britain in the global community do not express confidence in their understanding of and approach to issues. Rather than expressing willingness to encounter risk in seeking new sources or forms of personal power, these plays communicate a loss of personal power as new dangers engulf a formerly secure area of experience – the nation and its place in the world. Unconventional dramatic structure and uncertain endings signal the confusion surrounding conditions, cultures, and conflicts around the world. Moral questions posed by realistically constructed characters with stories of terrible suffering remain unanswered. Identity proves mutable in the context of international affairs and unreliable as a means of responding to global trends that are changing the forms of power and terms of participation in power. This group of plays poses the most serious challenges and offers the least clear and reassuring perspectives for contemporary Britain.

7
Political Theatre in an Era of Disengagement

Political theatre since 1995 has gone far beyond surviving as a remnant of the activist 1970s; it has expanded, developed in new ways, and become a force of renewal in British theatre by responding energetically to the conditions of the post-Thatcher era. Political plays have helped to define post-Thatcher politics through the issues and themes they have brought to visibility, and they have contributed to the continually evolving definition of theatre through devising new strategies to address and engage audiences. Contemporary political theatre acknowledges the basis of the current mood of disengagement by questioning and invalidating many of the ideas and practices that have constituted political activism. At the same time, it challenges indifference to politics by articulating issues of great immediacy in forms that signal their urgency. It contests apathy and cynical detachment through its example of engagement, and involves audiences in thinking about social and political issues without advocating particular ideologies to frame interpretation of those issues. The rich variety of new writing that has reinvigorated political theatre in Britain since 1995 offers a striking exhibition of current perspectives on public life and citizenship of the nation and world.

Contemporary political theatre contrasts with the work associated with the alternative theatre movement that originated in the 1970s. Contemporary drama expresses a mood of isolation and mourning rather than joyful optimism and the perspective of lone playwrights rather than collectives. By far the majority of recent political plays are performed in mainstream theatres rather than informal or improvised venues, such as the outdoor platforms and community centres of the alternative theatre movement. Contemporary playwrights, whether established or new, seek existing institutions and venues for their work, rather than creating companies or theatres. Performed in the National Theatre, the Royal

Court, or small theatres on the fringe, contemporary political plays thus address the conscious theatre audience rather than a more haphazard and perhaps more diverse one. Their quality must satisfy the mainstream audience, which seeks in the theatre a reflection of its own literacy and social power. The political content of the plays draws on mainstream concerns, such as the development of individuals within frameworks of family and history, the coherence and continuity of national identity, the rule of law and maintenance of justice, and the quality of political leadership. The interests in overthrowing traditional institutions of control, asserting the power of informal collectives, and creating new forms for public and private relationships that were common in the alternative theatre movement play little part in current political theatre.

The approaches to issues and problems evident in post-Thatcher political plays begin with a rejection of idealism. There are no utopias in these plays, and the language of idealism, as shown in Penhall's *Blue/Orange*, Churchill's *Far Away*, and Bean's *The Mentalists*, has been corrupted by the powerful. Only young and unsophisticated individuals, such as the title character of *My Name Is Rachel Corrie* and Mill in Harris's *Further from the Furthest Thing*, voice genuine ideals. The plays universally critique mercantilism, reject the values of the marketplace, and view the power of global corporations as a serious threat; however, they do not invoke the rhetoric of socialism or other anti-capitalist movements or suggest a better future through any political ideology. They do not examine or revise history in order to open the possibility of a more enlightened future, as did the 1970s alternative theatre. Those that portray history tend to emphasize the difficulty of understanding and interpreting past events. A few plays, such as Ravenhill's *Some Explicit Polaroids*, find hope in the idea of reconciling or transcending the political oppositions of the past. Hope for the future, such as it is, lies in dissolving political divisions and discarding current systems of political identification and action.

In place of idealism, contemporary political drama offers a pragmatic humanism. Many of the plays locate the primary meaning of human life in the connection to others in family or community. Appeals for change often take the Aristotelian form of evoking the audience's pity and fear, but shun catharsis and typically end on a note of uncertainty. Vividly compelling individual characters, rather than groups, arouse empathy as they demonstrate human dignity and worth in the face of oppression, pain, poverty, and disappointment. Individuals, rather than groups, are targeted with the plays' appeals to conscience. Contemporary plays suggest that individuals hold the potential for considerable power. Even those who have been victimized, as in Lavery's *Frozen* or Munro's *Iron*, show strength and resourcefulness in re-establishing their personal power of choice. The anarchic violence of the characters in absurdist fantasies by Butterworth, McDonagh, or Burke, the modest aspirations of 14-year-old Billy in Stephens's *Herons*, and the coming to terms with horror through personal writing, as seen in Linden's *I Have Before Me a Remarkable Document Given to Me by a Young Lady from Rwanda*, all exemplify individual power. Analyses of individual situations, however, inevitably expose larger power structures, as can be seen in Williams's *Sing Yer Heart Out for the Lads*, Dear's *Power*, and McDonagh's *Pillowman*, insidious systems of control as in Ravenhill's *Shopping and Fucking*, or global forms of domination as in *Gagarin Way*. Faced with such forces, any individual action appears futile, and these plays do not offer a means of reconciling individual conscience, aspirations, or action with oppressive systems of power. The limited power of the individual compared to that of economic and political systems immune to humanistic concerns thus forms a central unresolved issue in current political drama. Most current political plays thus call for a change in thinking rather than any form of action.

The in-yer-face plays first confronted the genteel apathy that had stifled political debate by bringing to public attention a generation in crisis. The spectacles of alienation, suffering, and destructiveness in plays by young dramatists testify to a tattered social fabric that has failed to protect children and youth or guide them towards autonomy. In-yer-face plays emphasize the consequent weakness and disability of characters representing the post-Thatcher generation. Their extreme situations and heightened emotion defy rational problem solving and create a catastrophic and unending present that would communicate hopelessness except for the energy evident in the interaction between the play and the audience. In place of hope, the in-yer-face plays offer a quality of immediacy and a sense of risk that engage the audience as they are confronted with scenes of cruelty, stark deprivation, human degradation, and pervasive amorality. While invalidating collective action and undermining basic vehicles of meaning such as identity and narrative, they explore the dark extremes of individual thought, action, and endurance. Though they make it clear that the odds are against surviving and that no hero's garlands await those who act, the plays show individual characters who sometimes find in themselves the power to break out of their restrictions. These characters model the dramatic choice of risking everything to make an individual change that may, in turn, transform the meaning or influence the course of events around them. The extremes evoked in the in-yer-face plays negate social or political action, but the emphasis on risk propels the audience towards the unknown and experimental. The intergenerational dialogue initiated by responses to the in-yer-face plays re-introduces themes of shared identity and history. These plays translate the alienation and energy of the post-Thatcher generation into critiques of society with less extreme conditions of risk and a more secure grounding in familiar situations. The responses emphasize change over the course of generations, whether for better or worse, and in that emphasis communicate the possibility of mobilizing energy to address concrete social and political issues. While they demonstrate the recognition that there will be no simple return to the pre-Thatcher political environment, the responses to the in-yer-face plays define the major political challenge of post-Thatcher Britain as rebuilding a communal sense of political power and purpose. They acknowledge an acute crisis among those most affected by major changes that began during the Thatcher era, such as the loss of skilled employment for the working class, the emergence of a deprived and marginalized underclass, and the failures of social and economic integration for Britain's cultural and racial minorities. These plays analyse the construction of individual and group identities as a means of moving towards a greater range of choice. They critique traditional institutions such as family and community, and urge attention to the breakdown of coherence and support in those institutions, but recognize and find hope in the human desire for connection and the continuing need to find meaning in connections with family and community.

Against the background of rapid and momentous change during the late twentieth century, the breakdown of traditional patterns, and the questioning of established political approaches, a group of playwrights has explored forms of power, analysing their operation and testing their resiliency. Plays exploring the power that resides in individuals and collectives, nations and religions, abstract structures such as knowledge or moral codes, and ideals such as loyalty show both oppressive institutions and moments when individuals dramatically exercise personal choice. These plays pose difficult questions about the relation of individuals to the structures of power. Works such as Frayn's *Democracy* and *Copenhagen* and Bennett's *The History Boys* complicate attempts to answer these questions by emphasizing the impurity of human motives, the instability of narrative, and the uncertainties in perceptions of reality. At the same time, by placing their questions of power in a historical context, they encourage an understanding of politics as an open-ended process in which different forms of power develop, change, and decline.

The enthusiastic response of audiences to political drama has empowered playwrights to articulate specific political issues and make implicit demands for change. Political drama has played an important part in making racism a visible issue in British society. The landmark tribunal play *The Colour of Justice* and Gupta's partly verbatim *Gladiator Games* expose tolerance of and collusion in racism by public officials and make implicit demands for reform. Political theatre keeps the conflict in Northern Ireland before the public, recognizing rights and wrongs on both sides while encouraging the reform of existing political entities as a move towards reconciliation. The contemporary issue that has generated the most new writing is the always highly visible one of political leadership. Playwrights have questioned or attacked the current leadership through satire, Brechtian analysis, and verbatim drama. Plays about current government leadership highlight a serious failure of trust, an atomization of responsibility that confounds accountability, and a centralizing tendency that ignores and often aggravates local problems. Though they do not indicate solutions to the problems they identify, the plays about specifically British issues exhibit a confidence and clarity communicated through topical references and recognizable or familiar characters. Response to these plays has been vigorously positive, and works such as Hare's *The Permanent Way*, seem to have prompted actual change in specific policies and practices.

Political theatre addresses the problems of global citizenship with less confidence and clarity than it brings to the purely national issues. The plays about international issues in which Britain has a stake show the greatest certainty when they deal with concerns that most directly involve the British nation. Thus, the verbatim play *Guantanamo*, about the detainees captured in Afghanistan and Iraq and imprisoned by the United States in Cuba, straightforwardly advocates release or speedy trial of the detainees, echoing widespread opposition among the British public to the Iraq war and the detentions associated with it. The plays that deal with the barriers faced by asylum seekers in the United Kingdom similarly argue for more sympathetic treatment of refugees and for changes to make the immigration process more just. When plays address issues that arise at a greater distance, however, they take a less sure tone, present the situations less directly, and often portray confusion in the response. That confusion and uncertainty characterize the plays like Gupta's *Sanctuary* or Waters's *World Music*, that deal with the African genocide. These plays suggest that British playwrights want to respond to events in Rwanda and other countries with tribal conflicts and violence but are uncertain how to present the conflicts and deal with questions of guilt and retribution. Issues such as terrorism and global business, which expose the limits of Western humanism, similarly elicit a confused response and inability to ask hard questions.

Current political theatre has developed new forms and used existing styles in novel ways to interpret the issues it has brought to public attention. The notable new developments are tribunal plays and verbatim plays. Tribunal plays, developed under the directorship of Nicolas Kent at the Tricycle Theatre, stage a hearing or trial that illuminates an important issue; an example is Norton-Taylor's *Justifying War*, which presents the enquiry into the death of David Kelly and focuses on the decision-making process through which Britain joined with the United States in invading Iraq. Verbatim plays, which have evolved primarily through the work of the Out of Joint Company directed by Max Stafford-Clark, use a variety of sources such as interviews, news reports, and speeches to construct a narrative tying together a series of events. Hare's *Permanent Way* exemplifies the verbatim play, while his *Stuff Happens* combines verbatim material with fictional dramatization. The factual material in tribunal and verbatim plays provides an antidote to distrust by presumably avoiding third-party interpretation (though, of course, selection of material involves a form of interpretation) and framing of issues in the rhetoric of particular political philosophies. More importantly, the tribunal plays substitute for failed public institutions and give audiences the opportunity of judging the issue, thus providing an immediate form of political engagement.

Contemporary political plays also use and combine a range of available styles to communicate the dimensions or urgency of their subjects. The extreme images of violence and physical pain in many of the in-yer-face plays can be associated with Artaud's Theatre of Cruelty. Edgar's *Playing with Fire* presents the central government's arrogance and short-sightedness through a series of Brechtian episodes. Brechtian structure characterizes Hare's *Permanent Way* and *Stuff Happens* as well. Churchill, DeAngelis, and McDonagh effectively use absurdism to heighten the sense of isolation and estrangement that their characters experience in the contexts of the breakdown of family and social support and the oppressive environments created by corporate or governmental power. Social realism allows understanding of the individual in a specific social setting in works by Khan-Din, Bean, Kwei-Armah, Munro, and others. Magical realism particularly characterizes the plays of Gupta, permitting her to show the character's realistic situation and metaphysical possibilities simultaneously.

The political theatre of contemporary Britain is a theatre of words. Most often it takes the form of scripted plays. Regardless of style, these plays use dialogue and language-based scenes to convey the specifics of characters, the social and political problems they present, and the themes they introduce. While experimentation with non-verbal forms of theatre does occur in Britain, it does not form a significant part of the techniques used by political dramatists. Most plays that successfully address political issues do so in terms that employ artfully constructed and polished language. Even the tribunal plays, which are taken from actual transcripts of hearings, often exhibit dazzling displays of wit or eloquence. This emphasis on the spoken word reinforces the focus on the power of the individual evident in other ways throughout the plays.

The current upsurge in political theatre continues even as this study of it concludes. While its future directions cannot be readily predicted, the body of new writing since 1995 brings fresh awareness of the potential of theatre as a public form for presenting and interpreting social and political issues. Political plays have proved capable of competing successfully for attention in a public deluged with information. When given the chance to engage directly with controversial issues, audiences have shown themselves eager to do so. Theatre has not demonstrated perfection: the absence of some important issues, the tentative explorations of others, and the occasional incidents of outright suppression represent failures of free and rigorous expression. Nevertheless, the stage remains a site of remarkable freedom and power with the potential to continually reinvent itself, present a full range of public issues, and create interpretations of these issues through artistic forms that engage the audience in constructing and exploring new political landscapes.

Notes

Chapter 1   Politics and Theatre

1 Shellard (1999: 12–13) discusses censorship of works by Unity Theatre, a socialist theatre company based in London, during the 1950s.

2 For further information about the theatre companies of the 1970s, see *Dreams and Deconstructions: Alternative Theater in Britain*, ed. Sandy Craig (Ambergate: Amber Lane Press, 1980) and *Stages in the Revolution: Political Theatre since 1960*, Catherine Itzin (London: Eyre Methuen, 1980).

3 Rees (1992) quotes John Ashford, the first editor of *Time Out*, as saying he introduced the term in the late1960s, borrowing the term that had come to categorize the non-invited performances at the Edinburgh Festival.

4 Eyre and Wright state, 'Arts Council funding for alternative theatres in 1969/ 1970 came to £15,000; ten years later, funding from the Drama Panel alone amounted to £1.5 million' (284).

5 Recent surveys have shown clear trends towards political disengagement and falling confidence in the efficacy of the political system. A well-known and accessible source of survey information is the yearly report *British Social Attitudes*, published in London by Sage Publications and the National Centre for Social Research. The 18th Report, edited by Alison Park *et al.* and published in 2001, included the information that 63% of respondents in 2000 thought the system of government needed improvement, compared to 49% in 1973. At the same time, 26% (compared with 19% in 1974) felt that parties 'are only interested in votes' rather than in the opinions and needs of constituents, and 19% (compared to 11% in 1991) felt that 'things go on the same' regardless of which party is in power. The same report shows only 6% of respondents with very strong political-party identification (compared to 11% in 1987), while 13% (compared to 8% in 1987) claim no party identification.

6 He also describes how rumours that he would not allow Catholics to be cast in his plays initially sabotaged an American production of his play *Trust* (5 April 2003).

7 See Wikipedia (www.wikipedia.org) for the history of the term 'feminazi'.

8 *Creative Britain*, the 1998 New Labour statement on arts policy authored by the Rt. Hon. Chris Smith, MP and Secretary of State for Culture, Media, and Sport, devoted full chapters to music, film, and broadcasting, but mentioned theatre in only 2 of its 170 pages.

9 Susan Bennett (1990) quotes a number of studies establishing a profile of the typical theatre audience as highly educated, affluent, and tending towards middle age. Initial studies of the results of a campaign within Britain to increase access to the arts indicate that the number of people who 'experience the arts regularly has surged by more than 800,000 since 2001' (*Guardian* 14 June 2004).

10 In a workshop I attended, Boal enquired about the occupation of participants and expressed some frustration over the narrowness of the self-selection process evident in the fact that everyone was either a teacher or a student.

11   Lyn Gardner, in 'Raising the Roof' (*Guardian* online edition 8 July 2002), remarks on a tremendous amount of cross-company collaboration.

Chapter 2   Generational Politics: The In-Yer-Face Plays

1   In 2001 the Royal Court Theatre presented its Sarah Kane Season, in which all her plays were revived.
2   See *General Household Survey*, 2000, issued by the Office for National Statistics and published in 2002, available on the national statistics website, www.statistics.gov.uk

Chapter 3   Intergenerational Dialogue

1   The *Guardian* columnist Madeleine Bunting recently suggested that the Blair government has 'cruised on Billy Elliot rhetoric', suggesting that increased opportunity that has enabled a few individual success stories has solved the problems of entire communities, while ignoring 'the friends, relatives, and neighbours left behind' (9–15 June: 5).

Chapter 4   Systems of Power

1   See, for example, Peter Riddell, *The Thatcher Era* (Oxford: Blackwell, 1991).
2   Speculation about the dynamics behind Brandt's resignation has included theories that he was set up by close associates who knew of Guillaume's spying, that he was seeking to avoid exposure of sexual misconduct, and that he was clinically depressed. See Hugh Eakin, 'On Stage and Off, the Mystery of Willy Brandt', *The New York Times* 18 December 2004: A19.
3   The playwright links the action of the play with speculation about secret US atomic testing in the South Atlantic in the late 1950s.

Chapter 5   Issues for Post-Thatcher Britain

1   The final report of the Inquiry recommended major changes in the organization, training, and practice of the Metropolitan Police as a result of the evidence that had been presented. For complete testimony, final report, and recommendations, see www.archive.official-documents.co.uk/document/cm42/4262/sli-03.htm.
2   Norton-Taylor had previously collaborated with John McGrath on *Half the Picture* in 1994, about hearings into arms deals with Iraq, and written *Nuremberg* in 1996, to commemorate the fiftieth anniversary of the Nazi war crimes tribunal.
3   Norton-Taylor, Richard, 'How Hutton Became a Play', *Guardian* 4 November 2003, online edition.
4   While noting that none of the shooting victims had been proved to be armed with either guns or bombs, the Widgery Inquiry gave credence to the

suspicion that some marchers had been firing at the army or supporting those who were firing. It also concluded that the shootings did not indicate a breakdown in military discipline, and avoided blaming individual soldiers for the deaths.

5 See Norton-Taylor, Richard, 'Bloody Sunday: The Final Reckoning Begins', *Guardian* 22 November 2004, online edition.

6 The soldiers who testified agreed to do so only on condition that they be questioned in London, not Londonderry, and that their identities not be revealed.

7 Reported in the *Guardian* 18 February 2003.

8 The Hutton Inquiry concluded that Kelly's death was a suicide and that allegations that the weapons dossier was altered to exaggerate the threat were not substantiated.

9 See, for example, 'Chronology of Rail Crashes' and 'Safety Crisis,' BBC News 10 May 2002, online edition.

10 David Hare noted, in an interview with *New York Times* reporter Elisabeth Bumiller, that his own admiration for Powell led him to portray him as something of a hero, but that he 'toughened up the writing about Powell' in response to knowledgeable critics who considered this too positive a view (*New York Times* 3 April 2006: A 14).

11 After the New York opening of the play, Colin Powell began to speak out about the prelude to war in Iraq, claiming that he knew there were no WMD and laying responsibility for the threats regarding such weapons on Vice President Cheney and the CIA. See Robert Scheer, 'Now Powell Tells Us', *The Nation* 26 April 2006, web edition.

12 *Market Boy* by David Eldridge, produced in 2006 at the National Theatre, made a personal and somewhat sentimental foray into the Thatcher years, but depicted Thatcher herself only as a larger-than-life gargoyle.

Chapter 6   Post-Thatcher Britain and Global Politics

1 Having enjoyed the novel experience of performing, Hare wrote a book about the experience, *Acting Up*, published by Faber and Faber in 1999.

2 The Refugee Council in the United Kingdom currently reports that asylum applications have fallen by 75% over the past three years (Refugee Council online).

3 In the original production this was done using a metal globe sculpture, for maximum symbolic resonance.

Bibliography

Plays

Adshead, Kay. *The Bogus Woman*. London: Oberon, 2001.

Bean, Richard. *The God Botherers*. London: Oberon, 2003.

Bean, Richard. *The Mentalists*. London: Oberon, 2002.

Bean, Richard. *Under the Whaleback*. London: Oberon, 2003.

Beaton, Alistair. *Feelgood*. London: Methuen, 2001.

Bennett, Alan. *The History Boys*. New York: Faber and Faber, 2004.

Bhatti, Gurpreet Kaur. *Behzti (Dishonour)*. London: Oberon, 2004.

Billy Elliot, the Musical. Programme. London: Victoria Palace, 2005.

Brenton, Howard. *Paul*. London: Nick Hern Books, 2005.

Brittain, Victoria and Gillian Slovo. *Guantanamo: Honor Bound to Defend Freedom*. London: Oberon Books, 2004.

Burke, Gregory. *Gagarin Way*. London: Faber and Faber, 2001.

Butterworth, Jez. *Mojo*. London: Nick Hern, 1998.

Churchill Caryl. *Blue Heart*. New York: Theatre Communications Group, 1997.

Churchill Caryl. *Far Away*. London: Nick Hern, 2000.

Churchill Caryl. *A Number*. London: Nick Hern, 2002.

Daley, Doña. *Blest Be the Tie*. London: Royal Court Theatre, 2004.

Darke, Nick. *The Riot*. London: Methuen, 1999.

De Angelis, April. *Wild East*. London: Faber and Faber, 2005.

Dear, Nick. *Power*. London: Faber and Faber, 2003.

Dunbar, Andrea and Robin Soans. *Rita, Sue, and Bob Too/A State Affair*. London: Methuen, 2000.

Dowie, Claire. *Easy Access (for the Boys)*. London: Methuen, 1998.

Edgar, David. *Playing with Fire*. London: Nick Hern, 2005.

Edmundson, Helen. *Mother Teresa Is Dead*. London: Nick Hern, 2002.

Eldridge, David. *Serving It Up* in *Plays: I*. London: Methuen, 2005.

Farr, David. *The UN Inspector*. London: Faber and Faber, 2005.

Feehily, Stella. *Duck*. London: Nick Hern, 2003.

Frayn, Michael. *Copenhagen*. New York: Anchor, 1998.

Frayn, Michael. *Democracy*. London: Faber and Faber, 2003.

Green, Debbie Tucker. *Born Bad*. London: Nick Hern, 2003.

Gupta, Tanika. *The Waiting Room*. London: Faber and Faber, 2000.

Gupta, Tanika. *Sanctuary*. London: Oberon, 2002.

Gupta, Tanika. *Gladiator Games*. London: Oberon, 2005.

Hare, David. *Via Dolorsa*. London: Faber and Faber, 1998.

Hare, David. *The Permanent Way*. London: Faber and Faber, 2003.

Hare, David. *Stuff Happens*. London: Faber and Faber, 2004.

Harris, Zinnie. *Further than the Furthest Thing*. London: Faber and Faber, 2000.

Kane, Sarah. *Blasted*, in *Complete Plays*, 1–61. London: Methuen, 2001.

Kane, Sarah. *Cleansed*, in *Complete Plays*, 105–151. London: Methuen, 2001.

Kane, Sarah. *Phaedra's Love*, in *Complete Plays*, 63–103. London: Methuen, 2001.

Kent, Nicolas. *Srebrenica*. London: Oberon Books, 2005.
Khan-Din, Ayub. *East Is East*. London: Nick Hern, 1996.
Kwei-Armah, Kwame. *Elmina's Kitchen*. London: Methuen, 2003.
Kwei-Armah, Kwame. *Fix Up*. London: Methuen, 2004.
Lavery, Bryony. *Frozen*. London: Faber and Faber, 2002.
Leigh, Mike. *Two Thousand Years*. London: Faber and Faber, 2006.
Linden, Sonja. *I Have Before Me a Remarkable Document Given to Me by a Young Lady from Rwanda*. London: Aurora Metro Publications, 2004.
Marber, Patrick. *Howard Katz*. London: Faber and Faber, 2001.
McDonagh, Martin. *The Beauty Queen of Leenane*, in *The Beauty Queen of Leenane and Other Plays*. New York: Vintage, 1998.
McDonagh, Martin. *The Lieutenant of Inishmore*. London: Methuen, 2001.
McDonagh, Martin. *The Pillowman*. London: Faber and Faber, 2003.
Mitchell, Gary. *Trust*. London: Nick Hern, 1999.
Mitchell, Gary. *The Force of Change*. London: Nick Hern, 2000.
Mitchell, Gary. *Loyal Women*. London: Nick Hern, 2003.
Morgan, Abi. *Skinned*. London: Faber and Faber, 1998.
Munro, Rona. *Iron*. London: Nick Hern, 2002.
Norton-Taylor, Richard. *The Colour of Justice: Based on the Transcripts of the Stephen Lawrence Inquiry*. London: Oberon, 1999.
Norton-Taylor, Richard. *Justifying War: Scenes from the Hutton Inquiry*. London: Oberon, 2003.
Norton-Taylor, Richard. *Bloody Sunday: Scenes from the Saville Inquiry*. London: Oberon, 2005.
Oglesby, Tamsin. *US and Them*. London: Faber and Faber, 2003.
Osborne, John. *Look Back in Anger*. New York: Bantam, 1965.
Penhall, Joe. *Some Voices* in *Plays: I*. London: Methuen, 1998.
Penhall, Joe. *Blue/Orange*. London: Methuen, 2000.
Pinnock, Winsome. *Mules*. London: Faber and Faber, 1996.
Prichard, Rebecca. *Yard Gal*. London: Faber and Faber, 1998.
Ravenhill, Mark. *Mother Clap's Molly House*. London: Methuen, 2001.
Ravenhill, Mark. *Shopping and Fucking*, in *Plays I*, 1–91. London: Methuen, 2001.
Ravenhill, Mark. *Some Explicit Polaroids*, in *Plays I*, 227–314. London: Methuen, 2001.
Rickman, Alan and Katharine Viner. *My Name is Rachel Corrie*. London: Nick Hern, 2006.
Soans, Robin. *Talking to Terrorists*. London: Oberon, 2005.
Stephens, Simon. *Herons*. London: Methuen, 2001.
Upton, Judy. *Ashes and Sand*, in *Plays: I*. London: Methuen, 2002.
Walsh, Enda. *Bedbound*. London: Nick Hern, 2000.
Waters, Steve. *World Music*. London: Nick Hern, 2003.
Wertenbaker, Timberlake. *Credible Witness*. London: Faber and Faber, 2001.
Williams, Roy. *Lift Off*. London: Methuen, 1999.
Williams, Roy. *Sing Yer Heart Out for the Lads*. London: Methuen, 2002.
Williams, Roy. *The No Boys Cricket Club, Plays: I*. London: Methuen, 2002.
Wynne, Michael. *The People Are Friendly*. London: Faber and Faber, 2002.

History and theory

Arendt, Hannah. *On Revolution*. New York: Viking, 1963.
Arts Council England. *Annual Review*, 2005, www.artscouncil.org.uk.

Aston, Elaine. 'The "Bogus Woman": Feminism and Asylum Theatre', *Modern Drama* 46:1 (Spring 2003) pp. 5–21.

Aston, Elaine. *Feminist Views on the English Stage*. Cambridge: Cambridge University Press, 2003.

'Asylum Facts and Figures', BBC online 23 July 2003.

Barker, Howard. *Arguments for a Theatre*. 3rd edn. Manchester: Manchester University Press, 1998.

Ben Chaim, Daphna. *Distance in the Theatre: The Aesthetics of Audience Response*. Ann Arbor, MI: UMI Press, 1984.

Benjamin, Walter D. *Selected Writings*, ed. Marcus Bullock and Michael W. Jennings. Cambridge, London: Harvard University Press, 1996. Vol. I: 1913–1926.

Benjamin, Walter D. 'Theses on the Philosophy of History', *Illuminations*, trans. Harry Zohn. New York: Schocken Books, 1969.

Bennett, Susan. *Theatre Audiences: A Theory of Production and Reception, 2nd edn.* London and New York: Routledge, 1997.

Billington, Michael. 'Drama Out of a Crisis', *Guardian* 10 April 2003, online edition.

Billington, Michael. 'Theatre of War', *Guardian* 17 February, 2001, online edition.

Blackadder, Neil. *Performing Opposition: Modern Theater and the Scandalized Audience*. Westport, CT: Praeger, 2003.

The Bloody Sunday Inquiry, www.bloody-sunday-inquiry.org/

Boal, Augusto. *Legislative Theatre: Using Performance to Make Politics*, trans. Adrian Jackson. London, New York: Routledge, 1998.

Brecht, Bertolt. *Brecht on Theatre: The Development of an Aesthetic*, ed. and trans. John Willet. New York: Hill and Wang, 1964.

Brook, Peter. *The Empty Space*. Avon, NY: Discus Books, 1969.

Bumiller, Elisabeth. 'Iraq War Faces Some New Critics (The Theater Kind)', *The New York Times* (3 April 2006) A14.

Bunting, Madeleine. 'Labour Just Isn't Listening'. *The Guardian Weekly* 174:25 (9–15 June 2006) 5.

Burke, Gregory. 'Funny Peculiar', *Guardian* 12 April 2003, online edition.

'Chronology of Rail Crashes' and 'Safety Crisis', BBC News 10 May 2002, online edition.

Deleuze, Gilles. 'One Less Manifesto: Theater and Its Critique', trans. Eliane dal Molin and Timothy Murray, in *Mimesis, Masochism, and Mime: The Politics of Theatricality in Contemporary French Thought*. Ann Arbor, MI: University of Michigan Press, 1997, 239–258. Originally published in *Anti-Oedipus, A Thousand Plateaus*, and *Kafka: Toward a Minor Literature*.

Dollimore, Jonathan. *Radical Tragedy: Religion, Ideology, and Power in the Dramas of Shakespeare and his Contemporaries*, 3rd edn. Basingstoke: Palgrave Macmillan, 2003.

Edelman, Murray. *Constructing the Political Spectacle*. Chicago and London: University of Chicago Press, 1988.

Edmunds, June and Bryan S. Turner, eds. *Generational Consciousness, Narrative, and Politics*. Lanham, MD: Rowman and Littlefield, 2002.

Edgar, David. 'Secret Lives', *The Guardian* 19 April 2003, online edition.

Edwardes, Jane, ed. *Frontline Intelligence 3: New Plays for the Nineties*. London: Methuen, 1995.

Eisenstadt, Shmuel. *From Generation to Generation*, 3rd ed. New Brunswick, NJ: Transaction Publishers, 2003.

Erickson, Jon. 'Defining Political Performance with Foucault and Habermas: Strategic and Communicative Action', *Theatricality*, eds. Tracy C. Davis and Thomas Postlewait. Cambridge: Cambridge University Press, 2003.

Eyre, Richard and Nicholas Wright. *Changing Stages: A View of British Theatre in the Twentieth Century*. London: Bloomsbury Publishing, 2000.

Fischer-Lichte, Erika. *History of European Drama and Theatre*. London: Routledge, 2002.

Foster, Hal, ed. *Postmodern Culture*. London: Pluto, 1985.

Gems, Pam. 'Not in their Name', *Guardian* 17 May 2003, online edition.

Goorney, Howard and Ewen MacColl. *Agit-prop to Theatre Workshop: Political Playscripts, 1930–50*. Manchester: Manchester University Press, 1986.

Griffin, Gabrielle. *Contemporary Black and Women Playwrights in Britain*. Cambridge: Cambridge University Press, 2003.

Grotowski, Jerzy. *Towards a Poor Theatre*. London: Methuen, 1975.

Hale, Edward Everett. *Dramatists of Today*. New York: Holt, 1905.

Hare, David. 'All Back to the Canteen', *Guardian* 24 May 2003, online edition.

Hobbes, Thomas. *Leviathan*, XIII, 1660.

The Hutton Inquiry, www.the-hutton-inquiry.org.uk.

Kattwinkel, Susan. *Audience Participation: Essays on Inclusion in Performance*. Westport, CT: Praeger, 2003.

Kershaw, Baz. *The Politics of Performance: Radical Theatre as Cultural Intervention*. London: Routledge, 1992.

Kershaw, Baz. *The Radical in Performance: Between Brecht and Baudrillard*. London: Routledge, 1999.

Klett, Renate. '"Anything's Possible in the Theatre": Portrait of Author Caryl Churchill'. *Theater Heute* (January 1984) p. 19.

Kustow, Michael. *theatre@risk*. London: Methuen, 2001.

Kwei-Armah, Kwame, 'Primary Colours', *Guardian* 10 May 2003, online edition.

Lacey, Stephen. *British Realist Theatre: The New Wave in Its Context, 1956–1965*. London: Routledge, 1995.

Lehmann, Hans-Thiess. *Postdramatic Theatre*, trans. Karen Jürs-Munby. London: Routledge, 2006.

McGrath, John. *Naked Thoughts that Roam About: Wrestling with Theatre, 1959–2001*. New York: Theatre Communications Group, 2002.

Meyrick, Julian. 'The Limits of Theory: Academic versus Professional Understanding of Theatre Problems'. *New Theatre Quarterly* 19:3 (August 2003) pp. 230–242.

Mitchell, Gary. 'Balancing Act', *Guardian* 5 April 2003, online edition.

Mullaney, Steven. *The Place of the Stage: License, Play, and Power in Renaissance England*. Ann Arbor, MI: University of Michigan Press, 1995.

Murray, Timothy, ed. *Mimesis, Masochism, and Mime: The Politics of Theatricality in Contemporary French Thought*. Ann Arbor, MI: University of Michigan Press, 1997.

Nicholson, Linda and Steven Seidman. *Social Postmodernism: Beyond Identity Politics*. Cambridge: Cambridge University Press, 1995.

Norton, Rictor. *Mother Clap's Molly House: The Gay Subculture in England, 1700–1830*. London: Gay Men's Press, 1992.

Norton-Taylor, Richard. 'How Hutton Became a Play', *Guardian* 4 November 2003, online edition.

Norton-Taylor, Richard. 'Bloody Sunday: The Final Reckoning Begins', *Guardian* 22 November 2004, online edition.

Olden, Mark. 'A Quick Chat with Ayub Khan-Din', 6 October 1999, www.kamera.co.uk.

Park, Allison *et al*. *British Social Attitudes*, vol. 18. London: Sage Publications and the National Centre for Social Research, 2001.

Patterson, Michael. *Strategies of Political Theatre: Post-War British Playwrights*. Cambridge: Cambridge University Press, 2003.

Peacock, D. Keith. *Radical Stages: Alternative History in Modern British Drama*. New York: Greenwood Press, 1991.

Peacock, D. Keith. *Thatcher's Theatre: British Theatre and Drama in the Eighties*. Westport, CT: Greenwood Press, 1999.

Ravenhill, Mark. 'A Tear in the Fabric: the James Bulger Murder and New Theatre Writing in the 'Nineties', *New Theatre Quarterly* 20:4 (November 2004) pp. 305–314.

Ravenhill, Mark. 'A Touch of Evil', *Guardian* 22 March 2003, online edition.

Rebellato, Dan. *1956 And All That: The Making of Modern British Drama*. London: Routledge, 1999.

Rebellato, Dan. 'Introduction', *Plays I*, by Mark Ravenhill, ix–xx. London: Methuen, 2001.

Riddell, Peter. *The Thatcher Era and its Legacy*. Oxford: Blackwell, 1991.

Russell, Richard Rankin. 'Loyal to the Truth': Gary Mitchell's Aesthetic Loyalism in *As the Beast Sleeps* and *The Force of Change*', *Modern Drama* VLVIII.1 (Spring 2005) pp. 186–209.

Saunders, Graham. 'Love Me or Kill Me': Sarah Kane and the Theatre of Extremes*. Manchester: Manchester University Press, 2002.

Schechner, Richard. *Essays on Performance Theory, 1970–1976*. New York: Drama Book Specialists, 1977.

Schechner, Richard. *Between Theatre and Anthropology*. Philadelphia, PA: University of Pennsylvania Press, 1985.

Scheer, Robert. 'Now Powell Tells Us', *The Nation* 26 April 2006, web edition.

Sierz, Aleks. *In-Yer-Face Theatre: British Drama Today*. London: Faber and Faber, 2000.

Sierz, Aleks. 'Still In-Yer-Face? Towards a Critique and a Summation', *New Theatre Quarterly* 18:69 (February 2002) pp. 17–24.

Smith, Chris. *Creative Britain*. London: Faber and Faber, 1998.

The Stephen Lawrence Inquiry, www.archive.official-documents.co.uk/document/cm42/4262/sli-03.htm.

Tallmer, Jerry. 'Jez Butterworth Discusses Latest Play at the Atlantic', *The Villager* 73:21 (24–30 September 2003) online edition.

Taylor, Paul. 'The Stage We're Going Through', *Independent* 10 June 2004, online edition.

Turner, Victor. *From Ritual to Theatre: The Human Seriousness of Play*. New York: Performing Arts Journal Publications, 1982.

Urban, Ken. 'Towards a Theory of Cruel Britannia: Coolness, Cruelty, and the 'Nineties', *New Theatre Quarterly* 20:4 (November 2004) pp. 354–372.

Wertenbaker, Timberlake. 'Dancing with History', *Crucible of Cultures: Anglophone Drama at the Dawn of a New Millenium*, ed. Marc Maufort and Franca Bellarsi. Brussels: P.I.E.-Peter Lang, 2002.

Wesker, Arnold. 'The Smaller Picture', *Guardian* 15 March 2003, online edition.

Williams, Raymond. *The Long Revolution*. New York: Harper and Row, 1966.

Wynne, Michael. 'Humour Me', *Guardian* 3 May 2003, online edition.

www.statistics.gov.uk. *General Household Survey, 2000*. Office for National Statistics, 2002.

www.statistics.gov.uk. 'UK Snapshot', continually updated, 2006.